THE TREATMENT OF COLLECTIVE COORDINATES IN MANY-BODY SYSTEMS

An Application of the BRST Invariance

World Scientific Lecture Notes in Physics

Published Titles

Vol. 25 : The Physics of Massive Neutrinos *by B Kayser with F Gibrat-Dehu and F Perrier*

Vol. 26 : QCD Spectral Sum Rules *by S Narison*

Vol. 27 : Covariant Operator Formalism of Gauge Theories and Quantum Gravity *by N Nakanishi and I Ojima*

Vol. 28 : Schrodinger's Mechanics *by D B Cook*

Vol. 29 : Introduction to Quantum Field Theory *by S-J Chang*

Vol. 30 : Integrable Models *by A Das*

Vol. 31 : Groups and Vector Spaces for Physicists *by C J Isham*

Vol. 32 : Modern Differential Geometry for Physicists *by C J Isham*

Vol. 33 : Special Relavitity *by V E Schroder*

World Scientific Lecture Notes in Physics Vol. 34

THE TREATMENT OF COLLECTIVE COORDINATES IN MANY-BODY SYSTEMS

An Application of the BRST Invariance

D R Bes and J Kurchan
CNEA, Buenos Aires, Argentina

World Scientific
Singapore • New Jersey • London • Hong Kong

Published by

World Scientific Publishing Co. Pte. Ltd.
5 Toh Tuck Link, Singapore 596224
USA office: 27 Warren Street, Suite 401-402, Hackensack, NJ 07601
UK office: 57 Shelton Street, Covent Garden, London WC2H 9HE

British Library Cataloguing-in-Publication Data
A catalogue record for this book is available from the British Library.

THE TREATMENT OF COLLECTIVE COORDINATES IN MANY-BODY SYSTEMS
An Application of the BRST Invariance

ISBN-13 978-981-02-0306-1
ISBN-10 981-02-0306-3
ISBN-13 978-981-02-0307-8 (pbk)
ISBN-10 981-02-0307-1 (pbk)

Contents

1 Introduction

"Why did I decide to add to the infinite
Series one more symbol? Why, to the vain
Skein which unwinds in eternity
Did I add another cause, effect, and woe?" J.L. Borges, The Golem [1]

The central object of these notes is the description of physical systems using variables which are subject to transformations, such as for example the coordinates measured with respect to a moving frame.

To the extent that the parameters determining the motion of the frame are just that (i.e., parameters with a time dependence which is externally determined), the resulting classical or quantal treatments do not bring in any essentially new feature. However, a different approach can be followed: these parameters may also be treated as genuine coordinates on an equal footing with the original ones. Since they affect the description of all the other variables they are traditionally called *collective coordinates*, while the remaining ones are called *intrinsic*. For instance, in the case of a rigid top, the quantal transformation from the laboratory to the rotating frame can be expressed in terms of Wigner's matrices $D^I_{MK}(\phi_v)$, depending on the Euler parameters ϕ_v. These parameters are treated as variables in the quantal treatment of the motion of the top, and the corresponding wavefunctions are given in terms of the same matrices D^I_{MK}.

This procedure may be generalized to cases in which the intrinsic variables of the system cannot be frozen (as in the case of the rigid top) and thus disregarded. Nuclei with large quadrupole moments behave approximately like quantum rotors. Superfluid systems display "rotational" bands associated with the number of pairs of particles. A similar situation arises when an extended classical solution of a field equation such as a soliton breaks the translational or rotational invariance. Such systems admit an (independent-particle) approximation that violates a symmetry of the hamiltonian (i.e., the approximate wavefunctions do not belong to an irreducible representation of the corresponding symmetry group). However, the exact wavefunctions do not break this symmetry (i.e., they can be labelled by the quantum numbers associated with the irreducible the representations). In such cases, the symmetry can be restored "through a motion in the broken symmetry group". At the level of small oscillations (or plane waves) the broken symmetries manifest themselves through the appearance of zero frequency modes (or massless fields) which make impossible the naive application of perturbative procedures. However, the zero modes do not yield a satisfactory description of the motion associated with the breaking of symmetries, if this breaking is only approximate (as for finite systems).

There are also other types of motion of many-body systems and extended objects which involve a great number of degrees of freedom and which can be associated with a certain group of transformations that are not necessarily symmetries (e.g. dilatations, quadrupole deformations, etc).

The most straightforward treatment that singles out these specially relevant degrees of freedom is to isolate them from the remaining variables. We may thus describe the system in a way that displays clearly the role played by the collective degrees of freedom (this, of course, does not mean that in general they will uncouple from the other variables). However, this procedure becomes unfeasible for realistic systems, since these cases must be treated within some approximation involving independent fermion and/or phonon fields, and in such representations it is very difficult to isolate the collective variables.

A second (less straightforward) approach is to use an overcomplete system of collective plus intrinsic variables. For example, in unified models in nuclear physics one has intrinsic coordinates describing the nucleons "in a rotating frame" plus collective variables describing the motion of the frame [2].

More precisely, throughout these notes we understand by collective coordinates as those describing a group of transformations on the system (in fact the main results of this work can be adapted to the more general case in which the set of transformations do not form a true Lie group). In general, we do not know how relevant these particular coordinates may be: this depends on the system under consideration. The choice of the transformation group is dictated by the intuition that we have a priori on the dynamics of the system. The case of strongly broken symmetries is specially clear since we know that these degrees of freedom dominate the low energy behaviour of physical systems.

The overcompleteness of variables however poses a problem already at a classical level, which becomes worse at the quantum level: it is not clear for example what is to be understood by "rotating frame", if the coordinates describing such a frame are quantum variables. The observation that clarifies this problem of overcompleteness is the following: *a classical system that is described in terms of variables subject to time-dependent (canonical) transformations and such that the parameters defining the transformations are treated as (collective) dynamical variables on an equal footing with the original variables possesses a local (i.e. time-dependent) gauge symmetry.* This gauge symmetry consists of the group of all time-dependent transformations which simultaneously moves the frame of reference and the particle system so as to reproduce the same physical situation. Choosing a particular transformation of the frame is, in this language, choosing a gauge.

Now, the classical as well as the quantum mechanical treatment of systems with gauge invariance has been extensively studied [3,4]. Following the previous observation and using the Faddeev-Popov [5] path-integral treatment of gauge systems, Gervais, Jevicki and Sakita and Hosoya and Kikkawa [6] formulated a functional version of the method of collective coordinates. This method was later adapted in ref. [7] for the hamiltonian treatment of the many-fermion problem. Several applications were made involving collective rotations of fermion systems in two and three dimensions [8]. However, the procedure as it stood was cumbersome to apply, specially in non-abelian cases. In order to simplify the calculations a different form of the method was implemented [9]. It was still based on the Faddeev-Popov path-integral but a gauge condition was imposed in a way that bears analogy with the Lorentz gauge in electromagnetism.

In recent years a simple and elegant approach for the treatment of gauge systems has called the attention of many field theoreticians. Instead of reducing the number of degrees of freedom in the overcomplete description, additional fermionic variables (ghosts) are introduced. This apparent complication simplifies the problem, because in gauge systems there is a supersymmetry involving variables and ghosts, which was discovered by Becchi, Rouet, Stora and Tyutin [10,11,12,13].

Recently we have showed how this method can be applied for the classical and quantal treatment of systems with collective coordinates [14]. The resulting formalism is purely algebraic and thus, in particular, it does not rely on functional methods (although path-integrals can readily be constructed a posteriori). We have applied this method to derive variational as well as perturbative approximations for rotating systems in two and three dimensions [15,16].

The outlay of these notes is as follows: the classical and quantal treatments of systems described from moving frame of references are discussed in sections 2 and 3, respectively. The central part of these notes is given in section 4, which contains the treatment following the (supersymmetric) invariance with respect to a BRST transformation. A parallel formulation in terms of path integrals (in fact the original formulation in [6]) is given in section 5. Broken symmetries are discussed in section 6. Sections 7 and 8 contain illustrative applications on abelian and non-abelian systems, respectively. The last section discusses a procedure to obtain (perturbatively) a purely collective dynamics. These notes admit different readings: the shortest path in order to operate with the formalism follows through sections 3, 4 and 6, and through the examples in sections 7 and 8. However, the notes are self-contained and thus they include the formalization and the proof of the BRST procedure. These aspects are emphasized in sections 2-5.

In these notes we intend to cover three main objectives. The first one is to present a formalism which both justifies and allows for the correction of the independent mean-field approximation in many-body problems displaying a breakdown of symmetries. Our second aim is to convey to the many-body physicists part of the beauty displayed in recent treatments of gauge systems. Finally, we illustrate these developments through simple mechanical examples.

Some items in these notes have not been published elsewhere, to the best of our knowledge. Such is the case, in particular, of the discussion concerning the structure and metric of the space of wavefunctions associated with the BRST treatment of collective coordinates.

Discussions with V. Alessandrini and F. Schaposnik have greatly stimulated this work. The support of colleagues and friends from TANDAR, C.N.E.A. is gratefully acknowledged.

2 The classical treatment of systems described by means of collective coordinates

The present section is devoted to the study of classical systems which are described from moving frames of reference. The parameters determining the motion of these frames are treated as dynamical variables on an equal footing with the original variables.

In subsection 2.1 we show the analogy between "gauging" a charged field to obtain electromagnetism and describing the motion of a particle in two dimensions from a rotating frame. The lagrangian and hamiltonian treatments of this mechanical system are discussed in subsection 2.2, stressing again the analogy with the electromagnetic case. In subsection 2.3 we consider the general case of a classical system whose dynamical variables are subject to an arbitrary (not necessarily symmetry) group of transformations. In appendix 2.A we discuss the particular problem of a general N-particle system which is described from a frame of reference that performs rotations in three dimensions. We thus exemplify some complications due to the non-abelian nature of the group of transformations. Finally, in appendix 2.B we derive some useful properties of Lie groups.

Our treatment makes use of the classical formalism for constrained systems developed by Dirac [17], as described in the presentation of Sudarshan and Mukunda [18].

2.1 The gauge nature of problems described from moving frames

Mechanical systems that are described from a moving frame of reference are indeed mechanical analogues of gauge field theories, with the simplification that the time is the single variable. As an example, we recall the derivation of the properties of the electromagnetic field using gauge methods [3] and, simultaneously, we translate this derivation to the case of a particle moving in a two-dimensional plane[1]. The expressions corresponding to the mechanical case are given below within square brackets.

We consider a complex scalar field $\Psi(x)$, where x denotes both space and time coordinates [$z(t) = q_1(t) + iq_2(t)$, where t is the time] whose classical lagrangian density [lagrangian] in the absence of electromagnetic coupling [in the laboratory frame] has the form

$$\mathcal{L}(\Psi(x), \partial_\mu \Psi(x)) = \frac{1}{2}\partial_\mu \Psi^* \partial^\mu \Psi - V(\Psi^*, \Psi) \tag{2.1}$$

$$L(z(t), d_t z(t)) = \frac{1}{2}|d_t z|^2 - V(z^*, z) \tag{2.2}$$

[1] Similar considerations can be found in [19] for the three dimensional case.

The gauge principle states that the electromagnetic coupling [the coupling between the intrinsic and rotational degrees of freedom] arises from an extension of "global invariance" to "local invariance". The term with derivatives in the lagrangian (2.1) [(2.2)] is obviously invariant under a constant phase change of Ψ [z]

$$\left.\begin{array}{lll}\Psi & \longrightarrow & U\Psi \\ z & \longrightarrow & Uz\end{array}\right\} U = exp[-i\alpha] \tag{2.3}$$

where α is an arbitrary real constant. This transformation is called a global (i.e. coordinate-independent) transformation and the theory is said to have global invariance under the group U(1). If also the potential has circular symmetry

$$V(\Psi^*, \Psi) = V(|\Psi|^2) \tag{2.4}$$

$$V(z^*, z) = V(|z|^2) \tag{2.5}$$

by Noether's theorem there is a conserved current

$$J_\mu = \Psi^* \overleftrightarrow{\partial}_\mu \Psi \tag{2.6}$$

$$J = -\frac{i}{2}(z^* \frac{dz}{dt} - z\frac{dz^*}{dt}) = q_1\dot{q}_2 - \dot{q}_1 q_2 \tag{2.7}$$

Now we consider local gauge transformations

$$\Psi \longrightarrow U(x)\Psi \quad U(x) = exp[-i\,\alpha(x)] \tag{2.8}$$

$$z \longrightarrow U(t)\,z \quad U(t) = exp[-i\,\alpha(t)] \tag{2.9}$$

where α is an arbitrary real function. That is, the transformations at different points of space-time [time] are independent of one another. Although $\partial_\mu\Psi$ [$d_t z$] transforms in the same manner as Ψ [z] under a global transformation, it acquires an extra term under a local transformation

$$\partial_\mu\Psi \longrightarrow U\partial_\mu\Psi + (\partial_\mu U)\Psi \tag{2.10}$$

$$d_t z \longrightarrow U d_t z + z d_t U \tag{2.11}$$

Therefore, the lagrangian is not invariant under a local transformation. In order to make the theory locally gauge invariant, we replace $\partial_\mu\Psi$ [$d_t z$] by a suitable generalization that transforms in the same manner as Ψ [z]. We define a vector [scalar] field $A_\mu(x)$ [$\omega(t)$] called a "gauge field", which transforms under the local transformation according to the rule

$$A_\mu(x) \longrightarrow A_\mu(x) + \frac{1}{e}\partial_\mu\alpha \tag{2.12}$$

$$\omega(t) \longrightarrow \omega(t) + d_t\alpha \tag{2.13}$$

Then, the "covariant derivative" ["invariant derivative"] defined by

$$D_\mu \Psi \equiv (\partial_\mu + ieA_\mu)\Psi \tag{2.14}$$

$$D_t z \equiv (d_t + i\omega)z \tag{2.15}$$

transforms in the same manner as Ψ [z]

$$D_\mu \Psi \longrightarrow U D_\mu \Psi \tag{2.16}$$

$$D_t z \longrightarrow (d_t + i\omega + i d_t \alpha) exp[-i\alpha] z = U D_t z \tag{2.17}$$

By replacing $\partial_\mu \Psi$ [$d_t z$] by $D_\mu \Psi$ [$D_t z$] in (2.1) [(2.2)], we obtain a new lagrangian density $\mathcal{L}(\Psi, D_\mu \Psi)$ [lagrangian $L(z, D_t z)$] which is obviously invariant under local gauge transformations. However, it contains the gauge field A_μ [ω] as an external field.

In Maxwell's theory one adds a term involving $\partial_\nu A_\mu$ quadratically. The only gauge-invariant Lorentz-scalar of this type is proportional to $F_{\mu\nu}F^{\mu\nu}$, where

$$F_{\mu\nu}(x) \equiv \partial_\mu A_\nu - \partial_\nu A_\mu \tag{2.18}$$

is called the field strength tensor. The lagrangian density for a closed dynamical system invariant under local gauge transformations is

$$\mathcal{L} = -\frac{1}{4} F_{\mu\nu}F^{\mu\nu} + \mathcal{L}(\Psi, D_\mu \Psi) \tag{2.19}$$

The factor $-\frac{1}{4}$ in the first term is purely conventional. The global U(1) symmetry, which is now enlarged with a local symmetry, is said to have been "gauged". The classical equations of motion following from (2.19) are just Maxwell's equations. Thus we have "derived" electromagnetism. The charge is a given real number that fixes the scale of A_μ relative to Ψ.

In the case of a two dimensional motion, the new lagrangian is

$$L = \frac{1}{2}|D_t z|^2 - V(|z|^2) = \frac{1}{2}[(\dot{q}_1 - \omega q_2)^2 + (\dot{q}_2 + \omega q_1)^2] - V(|z|^2) \tag{2.20}$$

The substitution of the invariant derivative in the lagrangian now yields the lagrangian in a rotating frame.

The lagrangian (2.20) may also be obtained through the replacement of q_i by[2] q_i' in the lagrangian (2.2), where

$$\begin{aligned} q_1' &= q_1 \cos\phi - q_2 \sin\phi \\ q_2' &= q_1 \sin\phi + q_2 \cos\phi \end{aligned} \tag{2.21}$$

Therefore, we identify the gauge field ω with the angular velocity $\dot{\phi}(t)$. If the time is our single variable, there is no quadratic term to be added, since

[2] From now on the prime and umprimed quantities refer to the laboratory and to the moving frame, respectively.

$F_{00} = 0$ by definition (eq.(2.18)). The inclusion of a charge would have no physical consequence other than interpreting ω as $\dot{\phi}/e$.

The lagrangian (2.20) is not merely invariant with respect to global (time-independent) transformation but it is also invariant with respect to local (time-dependent) ones. This is what is called gauge invariance. We notice that there appears an extra degree of freedom, the collective degree of freedom ϕ, in the lagrangian (2.20).

Since this example may be easily generalized, we conclude that *mechanical systems which are described from moving frames (and in which the parameters describing the motion of the frame are treated as genuine coordinates) are a particular type of gauge systems.* Consequently, in the quantisation of such problems we encounter many difficulties that are also present in the treatment of gauge theories. We intend to apply to the former the solutions that have been found for these last ones.

2.2 A particle moving in a two-dimensional plane

The lagrangian formalism

The Euler-Lagrange equations of motion

$$\frac{d}{dt}\left(\frac{\partial L}{\partial \dot{q}_i}\right) - \frac{\partial L}{\partial q_i} = 0 \tag{2.22}$$

can also be written as

$$\frac{\partial^2 L}{\partial \dot{q}_i \partial \dot{q}_j}\ddot{q}_j = \frac{\partial L}{\partial q_i} - \frac{\partial^2 L}{\partial \dot{q}_i \partial q_j}\dot{q}_j \tag{2.23}$$

Therefore, the accelerations at a given time are uniquely determined by the positions and velocities at that time if the determinant of the matrix $\partial^2 L/\partial \dot{q}_i \partial \dot{q}_j$ does not vanish. This is, for instance, the case for the lagrangian (2.2) in the laboratory frame

$$det\left(\frac{\partial^2 L}{\partial \dot{q}'_i \partial \dot{q}'_j}\right) = \begin{vmatrix} 1 & 0 \\ 0 & 1 \end{vmatrix} = 1 \tag{2.24}$$

However, we want to consider the lagrangian (2.20) corresponding to the description from a moving frame and to treat the *three* coordinates q_i ($q_3 \equiv \phi$) on an equal footing. In this case,

$$det\left(\frac{\partial^2 L}{\partial \dot{q}_i \partial \dot{q}_j}\right) = \begin{vmatrix} 1 & 0 & -q_2 \\ 0 & 1 & q_1 \\ -q_2 & q_1 & q_1^2 + q_2^2 \end{vmatrix} = 0 \tag{2.25}$$

Such a lagrangian is called singular. We want to show now that only two of the three equations of motion (2.23) are independent of each other. It may be straightforwardly verified that the vector $\vec{\eta}$ in the three dimensional space

$$\eta_1 = q_2 \quad ; \quad \eta_2 = -q_1 \quad ; \quad \eta_3 = 1 \tag{2.26}$$

satisfies the equation

$$\frac{\partial^2 L}{\partial \dot{q}_i \partial \dot{q}_j} \eta_j = 0 \tag{2.27}$$

Therefore, if we multiply (2.23) to the left by the vector $\vec{\eta}$, we can check that

$$\begin{aligned} \eta_i \left(-\frac{\partial^2 L}{\partial \dot{q}_i \partial q_j} \dot{q}_j + \frac{\partial L}{\partial q_i}\right) &= \frac{\partial V}{\partial q_1} q_2 - \frac{\partial V}{\partial q_2} q_1 + \frac{\partial V}{\partial q_3} \\ &= \frac{\partial V}{\partial q_1} \frac{\partial q_1}{\partial \phi}(q_i', \phi) + \frac{\partial V}{\partial q_2} \frac{\partial q_2}{\partial \phi}(q_i', \phi) + \frac{\partial V}{\partial \phi} \\ &= \frac{\partial V}{\partial \phi}\Big|_{q_1', q_2' = const.} = 0 \end{aligned} \tag{2.28}$$

where we have used the inverse of the transformation equations (2.21). This result expresses the obvious fact that the lagrangian in the laboratory frame does not depend on the coordinates of the moving frame. Therefore eq.(2.28), which is derived from the equations of motion, carries no dynamical information and is valid independently of the original form of V (in fact, it does not even require that the potential should be rotationally invariant). In this simple example we find that there is 'something missing', since we are not able to solve the equations of motion for the three variables. The solutions of the equations of motion of a singular lagrangian contain arbitrary functions of time. In our case, this is due to the fact that a lagrangian describing the system from a rotating frame lacks the information concerning the rotation of the frame itself (in electromagnetism, Maxwell's equations lack the information on the choice of the gauge).

As mentioned in subsection 2.1, the lagrangian (2.20) has a local time invariance which is independent of the original symmetries. It is invariant under the infinitesimal transformation

$$\begin{aligned} \delta q_1 &= \delta\alpha(t) q_2 \\ \delta q_2 &= -\delta\alpha(t) q_1 \\ \delta\phi &\equiv \delta q_3 = \delta\alpha(t) \end{aligned} \tag{2.29}$$

This invariance holds independently of the original form of the lagrangian in the laboratory frame. While the third eq.(2.29) changes the motion of the moving frame, the first two equations transform the coordinates q_1 and q_2 to the new frame. A gauge transformation maps a given trajectory into an equivalent one in which the same physical motion is described from another frame. Two trajectories belong to the same "class" if they are obtained from each other through a gauge transformation. All members of the same class represent the same physical trajectory. To choose a gauge means to select one member of each class. This is accomplished by imposing a supplementary relation between the coordinates and velocities.

$$G(q_i, \dot{q}_i, t) = 0 \tag{2.30}$$

G is called the gauge-fixing function and is arbitrary to a large extent. For instance, with the selection $G = \phi$ we are left with the original laboratory gauge. Another option is to choose $G = q_2$ which amounts to solve the problem in cylindrical variables, q_1 being the radial and ϕ the angular coordinate. Some other possibilities are discussed in following sections. It is precisely this freedom in the choice of G which makes a description using collective coordinates more flexible than the original one.

The hamiltonian formalism

The departing point of the hamiltonian formalism is to define the canonical momenta as functions of the coordinates and velocities. Our aim is to work in a phase-space $\mathcal{F}_2$ which includes ϕ, I as variables ($\phi \equiv q_3$; $I \equiv p_3$).

As usual, the conjugate momenta are defined by

$$p_i \equiv \frac{\partial L}{\partial \dot{q}_i} \qquad i = 1, 2, 3 \tag{2.31}$$

The vanishing of the determinant (2.25) is just the condition for the non-invertibility of the velocities as functions of the coordinates and momenta. The momenta are not all independent from each other: there exist relations between them which follow from (2.31) and which are called constraints. For instance, for the lagrangian (2.20), eq.(2.31) yields the three equations

$$\begin{aligned} p_1 &= \dot{q}_1 - q_2\dot{\phi} & (2.32)\\ p_2 &= \dot{q}_2 + q_1\dot{\phi} & (2.33)\\ I &= J & (2.34) \end{aligned}$$

where J is defined as in (2.7),

$$J = q'_1 \dot{q}'_2 - q'_2 \dot{q}'_1 = q_1 p_2 - q_2 p_1 \tag{2.35}$$

The two eqs. (2.32) and (2.33) can be considered to be definitions of momenta, while (2.34) is a constraint.

Since not all the velocities can be expressed in terms of the momenta, we choose the set of variables $(q_1, q_2, p_1, p_2, \phi, \dot{\phi})$ and we define the hamiltonian

$$H = \sum_{i=1}^{2} p_i \dot{q}_i + J\dot{\phi} - L$$

$$= \frac{1}{2}\sum_{i=1}^{2} p_i^2 + V \tag{2.36}$$

It is remarkable that the dependence on $\dot{q}_3$ disappears from the second line: this is a property of the Legendre transformation. We obtain the equation of motion following the usual procedure of varying both expressions for H with respect to the six variables and equating the coefficients of the independent variations [21]

$$\begin{aligned} \dot{p}_i &\simeq -\frac{\partial H}{\partial q_i} + \dot{\phi}\frac{\partial}{\partial q_i}(J - I) \\ \dot{q}_i &\simeq \frac{\partial H}{\partial p_i} - \dot{\phi}\frac{\partial}{\partial p_i}(J - I) \end{aligned} \tag{2.37}$$

where the symbol $\simeq$ denotes weak equality: equality when restricted to the constrained surface in $\mathcal{F}_2$ defined by (2.34). These equations are trivial for $\dot{q}_3 \equiv \dot{\phi}$. They imply that the time derivative of any phase space function M is[3]

$$\dot{M} \simeq \{M, H\} - \dot{\phi}\{M, (J - I)\} \tag{2.38}$$

where the Poisson brackets are those of $\mathcal{F}_2$. Notice that the constraint is preserved by the equations of motion. There is a dependence on the yet unspecified velocity $\dot{\phi}$, which is to be expected from the previous discussion using the lagrangian approach. Accordingly, we must choose a gauge fixing function $G(q_i, p_i)$. From eqs. (2.30) and (2.38)

$$0 = \dot{G} \simeq \{G, H\} - \dot{\phi}\{G, (J - I)\} \tag{2.39}$$

Solving $\dot{\phi}$ and introducing this value in eq. (2.34), we obtain the final expression for the equations of motion

$$\dot{M} \simeq \{M, H\} - \{M, (J - I)\}\frac{1}{\{G, (J - I)\}}\{G, H\} = \{M, H\}_D \tag{2.40}$$

The Dirac brackets

$$\begin{aligned} \{A, B\}_D \equiv \{A, B\} \quad &-\{A, (J - I)\}\frac{1}{\{G, (J - I)\}}\{G, B\} \\ &+\{B, (J - I)\}\frac{1}{\{G, (J - I)\}}\{G, A\} \end{aligned} \tag{2.41}$$

are a generalisation of the Poisson brackets and share most of their properties. From (2.40) we clearly see that if the constraint (2.34) and the gauge condition (2.30) are satisfied at $t = 0$, they will be satisfied at any subsequent time.

---

[3] We recall the definition of the Poisson bracket $\{A, B\} = \frac{\partial A}{\partial q_i}\frac{\partial B}{\partial p_i} - \frac{\partial A}{\partial p_i}\frac{\partial B}{\partial q_i}$

In **phase-space** $\mathcal{F}_2$, the finite transformation corresponding to the infinitesimal transformation (2.29) may be written

$$\begin{aligned} T(\alpha(t)) &= exp[-\alpha\{(J-I),\ \}] \\ &= 1-\ \alpha\{(J-I),\ \}+\frac{1}{2}\alpha^2\{(J-I),\{(J-I),\ \}\}+... \end{aligned} \tag{2.42}$$

Thus $(J-I)$ is the generator of the gauge transformation. The Poisson bracket between this generator and any physical magnitude vanishes, reflecting the fundamental invariance of gauge systems: while I changes the rotation of the frame of reference, J changes the coordinates in an appropriate way to leave the description invariant. In particular, the transformation (2.42) leaves invariant the phase-space lagrangian.

As in configuration space, the transformation (2.42) maps a given trajectory in phase space into another trajectory representing the same physical trajectory, but described from another gauge (i.e., from a different rotating system). This group of transformations is the gauge group. It leaves invariant the constraining hypersurface, which is mapped into itself.

Up to now we have seen that the original dynamics of the problem can be reproduced in a six-dimensional **phase-space** $\mathcal{F}_2$ with the constraint (2.34). We cast now these results in a form which is also used in the quantal treatments of sections 3 -5.

We impose the constraint by introducing a Lagrange multiplier Ω, which is also treated as a dynamical variable. The hamiltonian and phase-space lagrangian can be written in the new phase-space $\mathcal{F}_3$

$$\begin{aligned} H &= \frac{1}{2}\sum_{i=1}^{2}p_i^2+V-\Omega(J-I) \\ L &= \sum_{i=1}^{3}p_i\dot{q}_i+P\dot{\Omega}-H \end{aligned} \tag{2.43}$$

where we have introduced the canonical momentum P conjugate to Ω. The term $P\dot{\Omega}$ added in the lagrangian is zero if we impose the new constraint

$$P=0 \tag{2.44}$$

Hamilton's equation for P thus yields

$$0=\dot{P}=-\frac{\partial H}{\partial\Omega}=J-I \tag{2.45}$$

Therefore, the old constraint (2.34) appears now as a consequence of the equations of motion and hence it is called a secondary constraint, as opposed to eq.(2.44) which is a primary constraint (it is not a consequence of the dynamics).

In the present eight-dimensional phase-space $\mathcal{F}_8$ the gauge group of transformations is generated by

$$T(\beta(t)) = exp[-\dot{\beta}\{P, \ \} + \beta\{(J - I), \ \}] \tag{2.46}$$

which leaves invariant the lagrangian (2.43) in the moving frame when restricted to the surface defined by eq. (2.44).

To conclude this section we show that the hamiltonian H together with the primary and secondary constraints are a "caricature" of the electromagnetic problem. The electromagnetic field has the four-fold coordinates $A_\mu(x, t)$ where $\mu = i, t$ and $i = 1, 2, 3$ are the spatial indices. The lagrangian density is (c.f. eq.(2.19))

$$\mathcal{L} = -\frac{1}{4}F_{ij}F^{ij} - \frac{1}{2}F_{oi}F^{oi} \tag{2.47}$$

The momenta are defined in the usual way

$$E_\mu(x) = \frac{\partial \mathcal{L}}{\partial(\partial_o A_\mu)} = \dot{A}_\mu + \partial_\mu A^o \tag{2.48}$$

They represent the electric field components for $\mu = i$. For $\mu = 0$

$$E_o = \dot{A}_o + \partial_o A^o = 0 \tag{2.49}$$

which is a primary constraint. The hamiltonian density equals

$$\begin{aligned} \mathcal{H} &= \dot{A}_\mu E^\mu - \mathcal{L} = \frac{1}{2}E_i E^i + \frac{1}{4}F_{ij}F^{ij} - (\partial_i A^o)E^i \\ &\to \frac{1}{2}E_i E^i + \frac{1}{4}F_{ij}F^{ij} + A^o(\partial_i E^i) \end{aligned} \tag{2.50}$$

where we have added a divergency. From the equations of motion for E_0 we obtain

$$\dot{E}_o = \frac{\partial H}{\partial A_o} = \partial_i E^i(x) \tag{2.51}$$

Therefore the Gauss law appears as a secondary constraint. Table 1 shows the relation between the electromagnetic and the mechanical case. In electromagnetism, as in classical mechanics employing moving frames, we must impose additional gauge fixing conditions.

2.3 A general canonical transformation

In the present subsection we generalise the previous treatment of collective coordinates for two-dimensional rotations to the case of general canonical transformations[4].

[4]Although we sometimes continue to use the vocabulary associated with rotations, we are not restricted to this type of transformations.

	Electromagnetism	Mechanical case
Coordinates	$A_\mu(x,t)$	ϕ, Ω, q_1, q_2
Momenta	$E_\mu(x,t)$	I, P, p_1, p_2
Primary constraint	$E_o(x) = 0$	$P = 0$
Secondary constraint	$\partial_i E^i = 0$	$J - I = 0$

Table 1: The analogy between the electromagnetic and the mechanical case.

The laboratory variables in the original phase space $\mathcal{F}_1$ are denoted by q'_i, p'_i $(i = 1, ..., n)$. There are k generators which form a Lie algebra [5]

$$\{J_v, J_w\} = c_{vws} J_s \tag{2.52}$$

They generate the transformations

$$\tilde{T}(\phi_s) = exp\left[\phi_v \{J_v, \ \ \}\right] \tag{2.53}$$

where the ϕ_v are parameters[6]. Given a phase-space function $M'(q'_i, p'_i)$ of the original variables we reexpress it in terms of the transformed variables q_i, p_i as follows

$$M'(q'_i, p'_i) = T^{-1} M'(q_i, p_i) = M(q_i, p_i) \tag{2.54}$$

where $M'(q'_i, p'_i)$ and $M'(q_i, p_i)$ have the same functional dependence while $M'(q'_i, p'_i)$ and $M(q_i, p_i)$ take the same values. In particular, we can write a relation like (2.54) for the generators J_v themselves, and thus obtain a set of transformed generators which satisfy the same Poisson bracket algebra (2.52).

Since the transformation is an artifact introduced by us, the laboratory version of physical functions should be independent of the transformation parameters ϕ_v. This fact implies that not all the functions M are physical. In order to be so, they must satisfy the condition

$$\begin{aligned} 0 &= \tilde{T}^{-1} \frac{\partial}{\partial \phi_v} M'(q_i, p_i, \phi_w) = \tilde{T}^{-1}(\frac{\partial \tilde{T}}{\partial \phi_v}) M + \frac{\partial M}{\partial \phi_v} \\ &= \varsigma_{wv}(\phi_s)\{J_w, M\} + \frac{\partial M}{\partial \phi_v} = D_v M \end{aligned} \tag{2.55}$$

[5]For compact Lie groups the structure constants can be taken to be antisymmetric in the three indices v, w, s. We will assume this to be the case throughout these notes

[6]It may also be possible to parametrise the transformation in a different way, as with the Euler angles in the case of three-dimensional rotations.

The matrix $\varsigma(\phi_s)$ is a function of the parameters ϕ_s and depends on the particular group of transformations (see appendix 2.B). In eq. (2.55) we have introduced the "invariant" derivative

$$D_v \equiv \varsigma_{wv}\{J_w,\ \} + \frac{\partial}{\partial \phi_v} \tag{2.56}$$

We now raise the ϕ_v to the rank of true dynamical variables and introduce their associated momenta which satisfy

$$\{\phi_v, \mathcal{P}'_w\} = \{\phi_v, \mathcal{P}_w\} = \delta_{vw} \tag{2.57}$$

The $\mathcal{P}'$ and $\mathcal{P}$ complete the canonical set in the laboratory and in the transformed variables, respectively. We call this new phase-space $\mathcal{F}_2$. The definition of the transformation (2.53) must be slightly changed in order to insure that it continues to be canonical in the enlarged phase-space $\mathcal{F}_2$. Thus we now define, for instance in the canonical parametrization

$$T(\phi_s) = exp\left[\{\phi_v J_v,\ \}\right] \tag{2.58}$$

The transformation T coincides with the previous $\tilde{T}$ when applied to functions that do not contain the momenta $\mathcal{P}'_v$. However, we note that whereas the transformation $\tilde{T}$ in $\mathcal{F}_1$ is time-dependent (the time-dependency arising from the parameters ϕ_v) the transformation T in $\mathcal{F}_2$ is time-independent (the ϕ_v now being variables in $\mathcal{F}_2$). The transformation T can be verified to act on the collective momenta as follows

$$\mathcal{P}'_v = T^{-1}\mathcal{P}_v = -F_v \qquad F_v \equiv \varsigma_{wv} J_w - \mathcal{P}_v \tag{2.59}$$

Therefore, the conditions for a function to be physical read

$$\begin{aligned} \{M'(q'_i, p'_i, \phi_w), \mathcal{P}'_v\} &= 0 \\ \{M(q_i, p_i, \phi_w), F_v\} &= 0 \end{aligned} \tag{2.60}$$

in the laboratory and transformed frame, respectively.

We may also use an alternative form of these conditions. Introducing the matrix $\eta(\phi_s)$ that is inverse to ς, we obtain from the last of eqs. (2.60)

$$\begin{aligned} \{M, f_v\} &= 0 \\ f_v &\equiv \eta_{wv} F_w = J_v - I_v \end{aligned} \tag{2.61}$$

where the collective generators are defined as

$$I_v \equiv \eta_{wv} \mathcal{P}_w \tag{2.62}$$

In the laboratory frame the constraints f_v take a similar form

$$f'_v = -\eta_{wv} P'_w \tag{2.63}$$

The following commutation relations hold for the general case

$$\begin{aligned} \{J_v, J_w\} &= c_{vws} J_s \qquad \{I_v, I_w\} = -c_{vws} I_s \\ \{f_v, f_w\} &= c_{vws} f_s \qquad \{F_v, F_w\} = 0 \end{aligned} \tag{2.64}$$

These Poisson brackets are defined within the enlarged phase-space $\mathcal{F}_2$. The difference in sign for the collective angular momentum is well known since the pioneering work on rotation of molecules.

Since the collective coordinates ϕ_v, P'_v separate in the laboratory system (and in fact they are irrelevant for physical functions), we may impose for physical trajectories the conditions (slightly more severe than (2.55))

$$P'_v = 0 \tag{2.65}$$

which in terms of the transformed variables read

$$F_v = 0 \quad \text{or} \quad f_v = 0 \tag{2.66}$$

The Poisson brackets between two of these constraints and between the hamiltonian and each of them vanish on the hypersurface in $\mathcal{F}_2$ constrained by these equations. Thus, they are preserved at all times by the dynamics of the system.

In the original phase-space the lagrangian can be written

$$L' = p'_i \dot{q}'_i - H' = p'_i \dot{q}'_i + P'_v \dot{\phi}_v - H' - P'_v \dot{\phi}_v \tag{2.67}$$

The first three terms in the r.h.s. are the ones that one would expect from a lagrangian in $\mathcal{F}_2$. Since T is a time-independent canonical transformation in $\mathcal{F}_2$, we have (up to total time-derivatives)

$$p'_i \dot{q}'_i + P'_v \dot{\phi}_v - H' \to p_i \dot{q}_i + P_v \dot{\phi}_v - H \tag{2.68}$$

while from (2.59)

$$P'_v \dot{\phi}_v \longrightarrow -F_v \dot{\phi}_v \tag{2.69}$$

so that the phase-space lagrangian becomes

$$\begin{aligned} L &= p_i \dot{q}_i + P_v \dot{\phi}_v + F_v \dot{\phi}_v - H \\ &= p_i \dot{q}_i + \varsigma_{wv} J_w \dot{\phi}_v - H \\ &= p_i \dot{q}_i - (H - \omega_v J_v) \end{aligned} \tag{2.70}$$

where we have defined

$$\omega_v \equiv \varsigma_{vw}\dot{\phi}_w \tag{2.71}$$

which is a generalisation of the angular frequencies.

Following the same procedure as in (2.37) (i.e., varying H independently with respect to $p_i, q_i, \phi_v, \dot{\phi}_v$), we obtain

$$\dot{q}_i \simeq \frac{\partial H}{\partial p_i} - \dot{\phi}_v \hat{\varsigma}_{wv}\frac{\partial J_w}{\partial p_i} = \frac{\partial H}{\partial p_i} - \dot{\phi}_v \frac{\partial F_v}{\partial p_i} \tag{2.72}$$

$$\dot{p}_i \simeq -\frac{\partial H}{\partial q_i} - \dot{\phi}_v \hat{\varsigma}_{wv}\frac{\partial J_w}{\partial q_i} = -\frac{\partial H}{\partial q_i} + \dot{\phi}_v \frac{\partial F_v}{\partial q_i} \tag{2.73}$$

where again $\simeq$ denotes weak equality. Similar equations hold for ϕ_v, namely

$$\dot{\phi}_v \simeq -\dot{\phi}_w \frac{\partial F_w}{\partial \mathcal{P}_v} \tag{2.74}$$

which is trivial ($\dot{\phi}_v = \dot{\phi}_w \delta_{vw}$). From the requirement that the constraints are preserved at all times we obtain

$$\dot{\mathcal{P}}_w \simeq -\frac{\partial H}{\partial \phi_w} + \dot{\phi}_v \frac{\partial F_v}{\partial \phi_w} \tag{2.75}$$

so that we can write for any function in the $2n+2k$ phase-space $\mathcal{F}_2$

$$\dot{M} \simeq \{M, H\} - \dot{\phi}_v\{M, F_v\} = 0 \tag{2.76}$$

Note that if M is independent of $\phi_v, \mathcal{P}_v$, this equation reduces to the familiar expression (cf. eq. (2.70))

$$\dot{M} \simeq \{M, (H - \omega_v J_v)\} \tag{2.77}$$

Eq.(2.76) is a generalisation of eq.(2.37). It is as far as we can go without fixing a gauge.

Let us consider now k gauge-fixing functions G_v

$$G_v(q_i, p_i, \phi_s) = 0 \tag{2.78}$$

If these gauge conditions are to hold at all times

$$\dot{G}_v \simeq \{G_v, H\} - \dot{\phi}_w\{G_v, F_w\} = 0 \tag{2.79}$$

Provided that

$$det[\{G_v, F_w\}] \neq 0 \tag{2.80}$$

we may solve (2.79) for $\dot{\phi}_v$ and insert these derivatives in (2.76). We thus obtain the time evolution of an arbitrary function in terms of Dirac brackets

$$\begin{aligned}\dot{M} &\simeq \{M,H\} - \{M,F_w\}\ |\{G_\alpha,F_\beta\}|^{-1}_{wv}\ \{G_v,H\} \\ &\simeq \{M,H\} - \{M,f_w\}\ |\{G_\alpha,f_\beta\}|^{-1}_{wv}\ \{G_v,H\} \\ &\simeq \{M,H\}_D \end{aligned} \tag{2.81}$$

with

$$\begin{aligned}\{A,B\}_D \equiv\ & \{A,B\} - \{A,f_w\}\ |\{G_\alpha,f_\beta\}|^{-1}_{wv}\ \{G_v,B\} \\ & +\{B,f_w\}\ |\{G_\alpha,f_\beta\}|^{-1}_{wv}\ \{G_v,A\}\end{aligned} \tag{2.82}$$

We can also verify that the phase-space lagrangian (2.70) is left invariant by the group of gauge transformations

$$T(\alpha(t)) = exp[-\alpha_v(t)\{f_v,\ \}] \tag{2.83}$$

These results can be summarised by saying that within the extended phase-space $\mathcal{F}_2$, physical trajectories are restricted to a hypersurface defined by the constraining equations (2.66). On this hypersurface each trajectory is copied an infinite number of times. By choosing a gauge, we select a copy of each class of equivalent trajectories.

Finally, we can include the constraints as in (2.44).

$$\tilde{H} = H - \Omega_v f_v \tag{2.84}$$

with the primary constraints

$$P_v = 0 \tag{2.85}$$

where

$$\{\Omega_v, P_w\} = \delta_{vw} \tag{2.86}$$

In this new phase-space $\mathcal{F}_3$, eqs.(2.68) appear as secondary constraints

$$0 = \dot{P}_v = -\frac{\partial\tilde{H}}{\partial\Omega_v} = f_v \tag{2.87}$$

One can check that in $\mathcal{F}_3$ the gauge group is generated by the transformation [12]

$$T(\beta(t)) = exp[(-\dot{\beta}_s + \beta_v\Omega_w c_{svw})\{P_s,\ \} + \beta_v\{f_v,\ \}] \tag{2.88}$$

which leaves the phase-space version of the lagrangian invariant when restricted to the hypersurface (2.85).

appendix 2.A The three-dimensional rotation of a N-particl system

If the N-particle system constitutes a rigid rotor, a suitable description can be obtained from a rotating frame associated with the principal axis of the tensor of inertia. In this case, the useful coordinates are the collective coordinates determining the orientation of the moving frame. In this appendix we study the generalisation of this description to the case of a system which is not completely rigid, and thus the internal degrees of freedom have to be considered as well as the collective ones. The overcomplete set of variables must be treated in a self-consistent way.

We start with the lagrangian in the laboratory frame

$$L' = \sum_i \frac{m_i}{2} |\dot{\vec{r}_i}|^2 - V(\vec{r}_i) \tag{2.89}$$

where $i = 1, 2...N$ and $\vec{r}_i$ is the position vector of the i-th particle. The expression for the lagrangian in the rotating frame may be found in textbooks

$$L = \sum_i \frac{m_i}{2} |\dot{\vec{r}_i} + \vec{\omega} \wedge \vec{r}_i|^2 - V(\vec{r}_i, \phi_v) \tag{2.90}$$

The angles ϕ_v parametrise the motion of the rotating frame. The relation between the components ω_v of the angular frequencies and the time derivative of the angles ϕ_v is written

$$\omega_v = \varsigma_{vw}(\phi_s)\dot{\phi}_w \tag{2.91}$$

where the matrix $\varsigma(\phi_s)$ depends on the chosen parametrisation. As in the previous subsection we calculate the 3N+3 momenta associated with the coordinates $\vec{r}_i, \phi_v$ (which complete the new phase-space $\mathcal{F}_2$)

$$\vec{p}_i = m_i(\dot{\vec{r}_i} + \vec{\omega} \wedge \vec{r}_i) \tag{2.92}$$

$$\begin{aligned} P_v &= \sum_i m_i(\dot{\vec{r}_i} + \vec{\omega} \wedge \vec{r}_i)(\frac{\partial \vec{\omega}}{\partial \dot{\phi}_v} \wedge \vec{r}_i) \\ &= \sum_i m_i \frac{\partial \vec{\omega}}{\partial \dot{\phi}_v} \left[\vec{r}_i \wedge (\dot{\vec{r}_i} + \vec{\omega} \wedge \vec{r}_i)\right] = \frac{\partial \vec{\omega}}{\partial \dot{\phi}_v} \vec{J} \end{aligned} \tag{2.93}$$

where the generator $\vec{J}$ is defined

$$\vec{J} = \sum_i \vec{r}_i \wedge \vec{p}_i \tag{2.94}$$

We find again that the 3N+3 momenta are not independent of each other. They satisfy three constraint equations which play the same role as eq.(2.34)

$$F_v = \varsigma_{wv} J_w - P_v = 0 \tag{2.95}$$

A new set of constraints is obtained if we multiply F by η, (the inverse of ς).

$$f_v \equiv \eta_{wv} F_w = J_v - I_v = 0 \tag{2.96}$$

where

$$I_v \equiv \eta_{wv} P_w \tag{2.97}$$

are to be interpreted as the collective components of the angular momentum in the intrinsic frame. While the F_v commute between themselves, the J_v, I_v and f_v are generators of SU(2) algebrae (see appendix 2.B).

$$\begin{aligned} \{J_v, J_w\} &= \epsilon_{vws} J_s \qquad \{I_v, I_w\} = -\epsilon_{vws} I_s \\ \{f_v, f_w\} &= \epsilon_{vws} f_s \qquad \{F_v, F_w\} = 0 \end{aligned} \tag{2.98}$$

Here ϵ_{vws} is the Levi-Civita tensor.

Two useful parametrizations for this group are: i)the canonical parametrisation

$$T(\phi_s) = exp[\phi_v \tilde{J}_v] \tag{2.99}$$

and ii)the Euler parametrisation

$$T(\phi_s) = exp[\psi \tilde{J}_3]\, exp[\theta \tilde{J}_2]\, exp[\varphi \tilde{J}_3] \tag{2.100}$$

For the canonical parametrization we have

$$\eta_{vw} = \delta_{vw} \frac{\phi}{2} \cot\frac{\phi}{2} + \frac{\phi_v \phi_w}{\phi^2}\left(1 - \frac{\phi}{2}\cot\frac{\phi}{2}\right) - \epsilon_{vws}\frac{\phi_s}{2} \tag{2.101}$$

$$\varsigma_{vw} = \delta_{vw} \frac{sin\phi}{\phi} + \frac{\phi_v \phi_w}{\phi^2}\left(1 - \frac{\sin\phi}{\phi}\right) - \epsilon_{vws}\phi_s \frac{(1 - cos\phi)}{\phi^2} \tag{2.102}$$

with

$$\phi = \sqrt{\phi_v^2}$$

Similar relations for the Euler parametrization [20] are read off from (cf. eq. (2.97))

$$\begin{aligned} I_1 &= cos\psi\, P_\theta + sin\psi\, cosec\theta\, P_\varphi - cot\theta\, sin\psi\, P_\psi \\ I_2 &= -sin\psi\, P_\theta + cos\psi\, cosec\theta\, P_\varphi - cot\theta\, cos\psi\, P_\psi \\ I_3 &= P_\psi \end{aligned} \tag{2.103}$$

Similarly to the previous cases, we have three unspecified velocities which should be fixed by means of the three gauge-fixing conditions

$$G_v(\vec{r}_i, \vec{p}_i, \phi_s) = 0 \tag{2.104}$$

A possible choice associating the moving frame with the principal axis is to equate G_v to the non-diagonal components of the quadrupole tensor

$$G_v = Q_{v+1.v+2} = \sum_i m_i (r_{v+1})_i (r_{v+2})_i \tag{2.105}$$

appendix 2.B Some useful properties of Lie groups

In this appendix we derive some results that are used throughout the present notes.

Let J_v $(v = 1, ..., k)$ be k phase-space functions satisfying the Lie algebra (2.52) with Poisson brackets. We have the identities:

$$\begin{gathered} \{J_v, J_w\} = -\{J_w, J_v\} \\ \sum_{(vws)\ cyclic} \{\{J_v, J_w\}, J_s\} = 0 \end{gathered} \tag{2.106}$$

which imply

$$\begin{gathered} c_{vws} = -c_{wvs} \\ \sum_{(vws)\ cyclic} c_{vwt} c_{tsu} = 0 \end{gathered} \tag{2.107}$$

We shall further assume for simplicity that the c_{ijk} are totally antisymmetric[7].

We define

$$\tilde{J}_v \equiv \{J_v, \ \} \tag{2.108}$$

and from (2.107) we find that:

$$[\tilde{J}_v, \tilde{J}_w] = c_{vws} \tilde{J}_s \tag{2.109}$$

where $[\,,\,]$ denotes the commutator. With this algebra we can generate a group of canonical transformations parametrised by ϕ_v. For example, in the canonical parametrisation

$$T(\phi_s) = exp[\phi_v \tilde{J}_v] \tag{2.110}$$

From the group composition law

[7] This can always be assumed for compact groups.

$$
\begin{aligned}
T(\phi_v^{(c)}) &= T(\phi_v^{(b)})T(\phi_v^{(a)}) \\
\phi_v^{(c)} &= g_v(\phi_s^{(b)}, \phi_s^{(a)})
\end{aligned}
\tag{2.111}
$$

We define the matrix $\eta_{vw}(\phi_s^{(b)})$ and its inverse $\varsigma_{vw}(\phi_s^{(b)})$

$$
\begin{aligned}
\eta_{vw} &= \frac{\partial g_v}{\partial \phi_w^{(a)}}\Big|_{\phi_s^{(a)}=0} \\
\varsigma_{vs}\eta_{sw} &= \delta_{vw}
\end{aligned}
\tag{2.112}
$$

Let us consider that the $\phi_v^{(a)}$ are small increments $\delta\phi_{(v)}^{a}$. To first order

$$
\begin{aligned}
\delta\phi_v^{(c)} &= \phi_v^{(b)} + \delta\phi_v^{(b)} = \phi_v^{(b)} + \eta_{vw}\delta\phi_w^{(a)} \\
\delta\phi_v^{(a)} &= \varsigma_{vw}\delta\phi_w^{(b)}
\end{aligned}
\tag{2.113}
$$

After replacement in (2.111) we obtain

$$
T(\phi_v + \delta\phi_v) = T(\phi_v)T(\varsigma_{vw}\delta\phi_w) \tag{2.114}
$$

where we have dropped the supraindex b labelling *all* the parameters that appear in (2.114). We also obtain to first order

$$
\begin{aligned}
T^{-1}[T(\phi_v + \delta\phi_v) - T(\phi_v)] &= T^{-1}\frac{\partial T}{\partial \phi_v}\delta\phi_v \\
T(\varsigma_{wv}\delta\phi_v) - 1 &= \varsigma_{wv}\tilde{J}_w\delta\phi_v
\end{aligned}
\tag{2.115}
$$

Therefore, if we subtract T and multiply to the left by T^{-1} in both members of eq. (2.114) we obtain[8]

$$
T^{-1}\frac{\partial T}{\partial \phi_v} = \varsigma_{wv}\tilde{J}_w \tag{2.116}
$$

Furthermore, we obtain from $T.T^{-1} = 1$

$$
\frac{\partial T}{\partial \phi_v}T^{-1} + T\frac{\partial T^{-1}}{\partial \phi_v} = \frac{\partial T^{-1}}{\partial \phi_v}T + T^{-1}\frac{\partial T}{\partial \phi_v} = 0 \tag{2.117}
$$

[8]The formula (2.116) has been derived assuming a canonical parametrization but it is easy shown to be valid with any other (good) parametrization.

Let us now derive differential relations for the η, ς. From eq. (2.116) we obtain

$$\begin{aligned}
\frac{\partial T}{\partial \phi_v} &= T\varsigma_{wv}\tilde{J}_w \\
\frac{\partial^2 T}{\partial \phi_v \partial \phi_w} &= T\varsigma_{sw.v}\tilde{J}_s + \varsigma_{sw}\tilde{J}_s\frac{\partial T}{\partial \phi_v} \\
&= T(\varsigma_{sw.v}\tilde{J}_s + \varsigma_{sw}\tilde{J}_s\varsigma_{tv}\tilde{J}_t) \qquad (2.118)
\end{aligned}$$

where the comma denotes derivation. Similarly,

$$\frac{\partial^2 T}{\partial \phi_w \partial \phi_v} = T(\varsigma_{sv.w}\tilde{J}_s + \varsigma_{sv}\tilde{J}_s\varsigma_{tw}\tilde{J}_t) \qquad (2.119)$$

Subtracting and multiplying by T^{-1} to the left yields

$$(\varsigma_{uw.v} - \varsigma_{uv.w})\,\tilde{J}_u + \varsigma_{sw}\varsigma_{tv}[\tilde{J}_s, \tilde{J}_t] = 0 \qquad (2.120)$$

where we have swapped the indices in the second term of (2.119). Using the structure constants and swapping indices again

$$\varsigma_{uw.v} - \varsigma_{uv.w} + \varsigma_{sw}\varsigma_{tv}c_{stu} = 0 \qquad (2.121)$$

These relations imply other relations between the inverse η

$$\eta_{ws}\eta_{vt.w} - \eta_{wt}\eta_{vs.w} = -c_{tsu}\eta_{vu} \qquad (2.122)$$

The commutation relations

Using the previous formulae we can derive the Poisson brackets (2.64) as follows

$$\{I_v, I_w\} = \{\eta_{sv}P_s\ ,\ \eta_{tw}P_t\}$$

$$= \{\eta_{sv}, P_t\}\ \eta_{tw}\ P_s - \{\eta_{tw}, P_s\}\ \eta_{sv}\ P_t$$

$$= \eta_{sv.t}\ \eta_{tw}P_s - \eta_{tw.v}\ \eta_{sv}P_t$$

$$= (\eta_{sv.t}\ \eta_{tw} - \eta_{sw.t}\ \eta_{tv})P_s$$

$$= -c_{vws}\eta_{ts}calP_t = -c_{vws}I_s$$

where eq. (2.122) has been used. From this relation we immediately find that:

$$\{f_v, f_w\} = \{(J_v - I_v), (J_w - I_w)\} = c_{vws}(J_s - I_s) = c_{vws} f_s$$

We may also calculate

$$\{F_v, F_w\} = \{(\varsigma_{uv} J_u - P_v), (\varsigma_{tw} J_t - P_w)\}$$

$$= -\varsigma_{sv.w}\, J_s + \varsigma_{tw.v}\, J_w + \varsigma_{sv}\, \varsigma_{tw}\, \{J_s, J_t\}$$

$$= \varsigma_{sv}\, \varsigma_{tw}\, c_{stu} J_u - \varsigma_{sv,w}\, J_s + \varsigma_{tw.v}\, J_t$$

$$= (\varsigma_{uw.v} - \varsigma_{uv.w} - \varsigma_{tv}\, \varsigma_{sw}\, c_{stu})\, J_s = 0$$

where we have applied eq. (2.121).

3 The Hilbert spaces associated with collective coordinates

3.1 The constraints

In the present section we translate the classical results of section 2 to the quantal language. We start with the Hilbert space $\mathcal{H}_1$ of wavefunctions $\Psi'(q_i')$ ($i = 1, ..., n$) and we consider a general group of unitary transformations $T(\phi_v)$. Its generators $\hat{J}_v$ ($v = 1, ..., k$) satisfy a closed algebra.

$$[\hat{J}_v, \hat{J}_w] = ic_{vws}\hat{J}_s \tag{3.1}$$

where [,] denotes the commutation operation (c.f. eq. (2.52)). Most developments in this and the following sections can be straightforwardly extended to the case in which the transformations $T(\phi_v)$ do not constitute a Lie group (see comment in section 4), although in the present work we restrict ourselves to such a case.

The transformation operator reads, in the canonical parametrization[9]

$$\hat{T} = exp[-i\phi_v\hat{J}_v] \tag{3.2}$$

It relates the wavefunctions Ψ' and operators $\hat{O}'$ in the laboratory frame to those in the moving frame

$$\begin{aligned} |\Psi'> &= \hat{T}|\Psi> \\ \hat{O}' &= \hat{T}\hat{O}\hat{T}^\dagger \end{aligned} \tag{3.3}$$

In particular, one obtains the transformed versions of the $\hat{J}_v$ themselves.

We wish to treat the k parameters ϕ_v as genuine coordinates. Therefore, we define the overcomplete Hilbert space $\mathcal{H}_2$ whose wave functions depend both on q_i and ϕ_v. We also introduce the operators $\hat{\phi}_v$ and their canonical momenta $\hat{P}_v$. We assume that $-\infty < \phi_v < \infty$ and thus that $\hat{P}_v$ has continuous eigenvalues[10]

$$\begin{aligned} [\hat{\phi}_v, \hat{P}_w] &= i\delta_{vw} \\ \hat{P}_v &= -i\frac{\partial}{\partial\phi_v} \end{aligned} \tag{3.4}$$

In $\mathcal{H}_2$, the laboratory version of physical states $|ph'>$ is independent of the ϕ_v (c.f. eq.(2.55)). (In fact they are the wavefunctions belonging to $\mathcal{H}_1$ times a unit state in the collective space).

[9]Any other parametrization of the transformation (such as the Euler parametrization of the three dimensional rotations) can be used as well.

[10]However, for example in the case of rotations, we will be free to work in the subspace of integer values of $\hat{P}_v$ since the rotational one-turn operator $exp[2\pi i\hat{P}_v]$ is, in practical cases a conserved magnitude, and thus we may choose periodic wavefunctions in ϕ_v.

$$
\begin{aligned}
0 &= i\hat{T}^{\dagger}\frac{\partial}{\partial\phi_v}|ph'>= \hat{F}_v|ph> \\
\hat{F}_v &\equiv i\hat{T}^{\dagger}(\frac{\partial\hat{T}}{\partial\phi_v})+i\frac{\partial}{\partial\phi_v} = -\hat{T}^{\dagger}\hat{P}_v\hat{T} = \varsigma_{wv}(\phi_s)\hat{J}_w - \hat{P}_v
\end{aligned} \tag{3.5}
$$

The alternative set of constraints is obtained through the multiplication of F by η (the inverse matrix of ς).

$$
\begin{aligned}
\hat{f}_v|ph> &= 0 \\
\hat{f}_v &\equiv \eta_{wv}F_w = \hat{J}_v - \hat{I}_v
\end{aligned} \tag{3.6}
$$

where

$$
\hat{I}_v \equiv \eta_{wv}\hat{P}_w \tag{3.7}
$$

Thus $\hat{I}_v$ is the collective version of $\hat{J}_v$. In addition to (3.1), the following commutation relations are satisfied [11].

$$
[\hat{P}_v, \hat{P}_w] = 0 \quad [\hat{I}_v, \hat{I}_v] = -ic_{vws}\hat{I}_s
$$

$$
[\hat{F}_v, \hat{F}_w] = 0 \quad [\hat{f}_v, \hat{f}_w] = ic_{vws}\hat{f}_s, \tag{3.8}
$$

The constraint operators are generators of the gauge group of transformations. Physical states carry the zero-representation of such group. Of course, this does not imply that they are left invariant by the transformations generated by the $\hat{J}_v$ or by the collective transformations generated by the operators $\hat{I}_v$ separately.

In practical cases it is easier to work with the constraints f_v. However, since the constraints F_v are in involution (i.e., they commute among themselves) they are useful for demonstration purposes (cf., for instance, subsection 4.2). For any physical operator acting on $\mathcal{H}_2$ (i.e., an operator that maps physical states into physical states) it can be seen that

$$
[\hat{O}_{ph}, \hat{F}_v] \simeq 0 \tag{3.9}
$$

$$
[\hat{O}_{ph}, \hat{f}_v] \simeq 0 \tag{3.10}
$$

The symbol $\simeq$ means that the l.h.s. vanishes if applied to physical states.

As in the classical treatment, we can impose the constraints (3.6) by further enlarging the Hilbert space as follows: we add k new variables Ω_v, their associated momenta P_v and the corresponding operators $\hat{\Omega}_v$ and $\hat{P}_v$ The constraints (3.6) are dynamically satisfied by the new hamiltonian $\hat{\hat{H}}$

$$
\hat{\hat{H}} \simeq \hat{H} - \Omega_v\hat{f}_v \tag{3.11}
$$

[11]The minus sign in the commutation relations for the I_v is obtained through a quantal transcription of the procedure followed in appendix 2.B

with new primary constraints

$$\hat{P}_v|ph>= -i\frac{\partial}{\partial\Omega_v}|ph>=0 \tag{3.12}$$

where the functions in the enlarged Hilbert space $\mathcal{H}_3$ depend on q_i, ϕ_v and Ω_v. The new hamiltonian preserves the original dynamics for physical states (i.e., for those states that satisfy the constraint (3.6)). Physical operators in $\mathcal{H}_3$ also satisfy

$$[\hat{O}_{ph}, \hat{P}_v] \simeq 0 \tag{3.13}$$

The Hilbert spaces $\mathcal{H}_1$, $\mathcal{H}_2$ and $\mathcal{H}_3$ are seen to be the quantal counterpart of the phase spaces $\mathcal{F}_1$, $\mathcal{F}_2$ and $\mathcal{F}_3$ (cf. subsection 2.3). The constraints (3.5), (3.6) and (3.12), are also the quantal versions of the constraints (2.66) and (2.85). They tell us that only a subspace of the enlarged Hilbert spaces $\mathcal{H}_2$ and $\mathcal{H}_3$ is relevant, the one annihilated by them. In appendix 3.B and section 4 we show two different formalisms that enable us to operate in these spaces. For this purpose we must impose the quantal versions of the gauge conditions (2.78).

Finally, it is useful to rewrite these results in the laboratory frame using the transformation (3.3). The laboratory version of the constraints (2.66) and (2.85) read:

$$\begin{aligned}
\hat{F}'_v|ph'>= -\hat{P}'_v|ph'>= i\frac{\partial}{\partial\phi_v}|ph'>=0 \\
\hat{f}'_v|ph'>= -\eta_{wv}\hat{P}'_w|ph'>= i\eta_{wv}\frac{\partial}{\partial\phi_w}|ph'>=0 \\
\hat{P}_v|ph'>= -i\frac{\partial}{\partial\Omega_v}|ph'>=0
\end{aligned} \tag{3.14}$$

whose meaning is clear: in the laboratory frame the physical wave functions of $\mathcal{H}_3$ are independent of the variables ϕ_v, Ω_v.

Eqs. (3.10) and (3.13) take the form:

$$[\hat{O}'_{ph}, \hat{F}'_v] \simeq [\hat{O}'_{ph}, \hat{f}'_v] \simeq i\frac{\partial}{\partial\phi_v}\hat{O}'_{ph} \simeq 0 \tag{3.15}$$

$$[\hat{O}'_{ph}, \hat{P}'_v] \simeq -i\frac{\partial}{\partial\Omega_v}\hat{O}'_{ph} \simeq 0 \tag{3.16}$$

3.2 The inner products

The inner product associated with the original variables q'_i in the Hilbert space $\mathcal{H}_1$ is given ab initio and we are not free to alter it. However, when we extend the space $\mathcal{H}_1$ to construct $\mathcal{H}_2$ and $\mathcal{H}_3$, we are introducing artificial dependencies of wavefunctions on the parameters ϕ_v, Ω_v. The definition of the inner products in $\mathcal{H}_2$ and $\mathcal{H}_3$ is to a large extent arbitrary, and we only need to require that for physical functions (those which satisfy the constraints) the overlaps coincide with the original ones in $\mathcal{H}_1$.

The measure is most easily constructed in the laboratory frame, in which it takes the product form. Given the overlap of two functions Ψ'_1, Ψ'_2 in the original space $\mathcal{H}_1$

$$\int \Psi_1^{*'} \; \Psi'_2 \; d\mu(q_i) \tag{3.17}$$

we extend it to functions in $\mathcal{H}_2$ in the product form

$$\int \Psi_1^{*'} \; \Psi'_2 \; d\mu(q_i) \; d\mu_\phi \tag{3.18}$$

In order to define the measure $d\mu_\phi$ we also require: i) that the operators $\hat{\phi}_v$ be hermitean (and thus that the transformation (3.2) is unitary) and ii) that the constraint operators be hermitean. This can be enforced either for the constraints $\hat{F}_v$ (eq.(3.5)) or for the $\hat{f}_v$ (eq.((3.6))). Thus the measure $d\mu_\phi$ associated with the collective coordinates ϕ_v depends on the choice of the set of constraints.

For the constraints $\hat{F}_v$ we define the measure as the simple product [12]

$$d\mu_\phi^{(F)} = \Pi_v d\phi_v \tag{3.19}$$

With this definition the operators $\hat{\phi}_v$, $\hat{P}_v$ and $\hat{F}_v$ are hermitean.

In the case of the constraints $\hat{f}_v$, we define the measure following the usual prescription for the rotational case [13]

$$d\mu_\phi^{(f)} = |\varsigma| \; \Pi_v d\phi_v \tag{3.20}$$

where $|\varsigma|$ denotes the determinant $det[\varsigma_{vw}]$. We show in appendix 3.A that the $\hat{f}_v$ (or, equivalently, the $\hat{I}_v$) are hermitean with respect to this measure.

At this point we notice that physical wavefunctions in $\mathcal{H}_2$ (which are independent of ϕ_v) have infinite norm. This is characteristic of gauge invariant systems. The problem can be cured within $\mathcal{H}_2$ by redefining the overlaps suitably. We outline this procedure in appendix 3.B. It will take us close to the treatment in the "Coulomb gauge" of section 5. However in these notes we emphasize the procedure discussed in section 4, where we proceed by further enlarging the Hilbert space.

We now go on to define the overlap associated with the coordinates Ω_v. Since the operators $\hat{P}_v$ commute among themselves (as well as with the constraint

[12] In these inner product, the integrals extend from $-\infty$ to ∞ (cf. the footnote on p.25).

[13] In the Euler parametrization of the rotational case

$$|\varsigma| = sin\beta$$

is the familiar factor appearing in the measure associated with the Wigner functions D^I_{MK} [20]

operators $\hat{F}_v$ and $\hat{f}_v$) we discuss first the case of a single coordinate Ω. We construct the operators

$$\Gamma_\Omega = \frac{1}{\sqrt{2}}(\hat{\Omega}+i\hat{P}) = \frac{1}{\sqrt{2}}(\Omega+\frac{\partial}{\partial\Omega})$$
$$\Gamma^\dagger_\Omega = \frac{1}{\sqrt{2}}(\hat{\Omega}-i\hat{P}) = \frac{1}{\sqrt{2}}(\Omega-\frac{\partial}{\partial\Omega}) \quad (3.21)$$

$$[\Gamma_\Omega,\Gamma^\dagger_\Omega] = 1 \quad (3.22)$$

We could construct a basis for the Hilbert space of the form

$$\psi_n^{(h.o.)} = \frac{1}{\sqrt{n!}}(\Gamma^\dagger_\Omega)^n\psi_o^{(h.o.)}$$
$$\Gamma_\Omega\psi_o^{(h.o.)} = 0$$
$$<\psi_n^{(h.o.)}|\psi_m^{(h.o.)}> = \delta_{nm} \quad (3.23)$$

This basis corresponds to the usual inner product in configuration space

$$<g|h>=\int_{\Omega=-\infty}^{\Omega=\infty} g^*(\Omega)h(\Omega)d\Omega \quad (3.24)$$

According to ref. [23] it is possible to use such a (positively defined) metric in connection with these coordinates.

However, using the freedom that we have in the definition of operations in the unphysical space, we may introduce an indefinite metric already at this level. It turns out that this definition is very convenient both from the theoretical point of view (an equivalence is proved straightforwardly without problems of regularization) and from the practical point of view (the actual calculations are indeed simpler). We discuss this in further detail in the next section.

The rationale is that the "error" spuriously produced by enlarging the Hilbert space with the collective coordinate ϕ may be "cancelled" by the introduction of the Lagrange multiplier Ω[14].

$$|n_\Omega> = \frac{1}{\sqrt{n!}}(\Gamma_\Omega)^n|0_\Omega>$$
$$\Gamma^\dagger_\Omega|0_\Omega> = 0$$
$$<n_\Omega|m_\Omega> = (-1)^n\delta_{nm} \quad (3.25)$$

Hence $\mathcal{H}_3$ has indefinite norm and is not a "Hilbert" but a "pseudo-Hilbert" space. Let us first explicitly construct this basis in the coordinate representation. The second of eqs. (3.25) implies

$$|0_\Omega>= exp[\Omega^2/2] \quad (3.26)$$

[14]This is the case in electromagnetism where an indefinite metric for the photon associated with the Lagrange multiplier is used [22].

and thus

$$|n_\Omega>=\frac{1}{\sqrt{n!}}(\Omega+\frac{\partial}{\partial\Omega})^n|0_\Omega>=(-i)^n\psi_n^{(h.o.)}(i\Omega) \tag{3.27}$$

The inner product associated with (3.25) can be written, in coordinate representation,

$$\begin{aligned}<g|h> &= -i\int_{\Omega=-i\infty}^{\Omega=i\infty}[g(-\Omega)]^*h(\Omega)d\Omega \\ &= \int_{\tau=-\infty}^{\tau=\infty}[g(-i\tau)]^*h(i\tau)d\tau \end{aligned} \tag{3.28}$$

where $\Omega = i\tau$ and the conjugation applies both to the constants defining the function g and to the argument. In (3.28) Ω should be regarded as imaginary and τ as real, since the integration path is taken to be the imaginary axis in Ω. It is straightforward to show that this inner product satisfies the requirement that $\hat{\Omega}$ and $\hat{P}$ should be hermitean. This measure is implicitly used in refs. [9,15,16].

Finally, we obtain for the inner product of wavefunctions in $\mathcal{H}_3$ (laboratory frame)

$$\begin{aligned}<\Psi_1'|\Psi_2'>_{\mathcal{H}_3}^{(F)} &= (-i)^k\int_{\Omega_v=-i\infty\ \phi_v=-\infty}^{\Omega_v=i\infty\ \phi_v=\infty}\Pi_v\ d\Omega_v\ d\phi_v\ d\mu(q_i') \\ &\quad [\Psi_1'(-\Omega_w,\phi_w,q_i')]^*\Psi_1'(\Omega_w,\phi_w,q_i')\end{aligned} \tag{3.29}$$

$$\begin{aligned}<\Psi_1'|\Psi_2'>_{\mathcal{H}_3}^{(f)} &= (-i)^k\int_{\Omega_v=-i\infty\ \phi_v=-\infty}^{\Omega_v=i\infty\ \phi_v=\infty}|\varsigma|\ \Pi_v\ d\Omega_v\ d\phi_v\ d\mu(q_i') \\ &\quad [\Psi_1'(-\Omega_w,\phi_w,q_i')]^*\Psi_1'(\Omega_w,\phi_w,q_i')\end{aligned} \tag{3.30}$$

With the overlaps $<\ |\ >_{\mathcal{H}_3}^{(F)}$ and $<\ |\ >_{\mathcal{H}_3}^{(f)}$ the constraints $\hat{F}_v$ and $\hat{f}_v$ are hermitean, respectively.

The problem of infinities in the overlaps of physical wavefunctions is solved in the next section through a further enlargement of the space.

appendix 3.A The hermiticity properties of operators using the measure associated with the constraints f_v

To prove the hermiticity of f_v we must show that

$$\int h^{*'}(\eta_{wv}\frac{\partial}{\partial\phi_w})g'\, d\mu_\phi^{(f)} = -(\int g^{*'}(\eta_{wv}\frac{\partial}{\partial\phi_w})h'\, d\mu_\phi^{(f)})^* \tag{3.31}$$

for every g', h' in the lab frame (cf. the second of eqs. (3.14)). Assuming that surface terms vanish, eq.(3.31) requires

$$\frac{\partial}{\partial\phi_w}(|\varsigma|\ \eta_{wv}) = 0 \tag{3.32}$$

Eq. (3.32) is equivalent to

$$\begin{aligned} 0 &= \eta_{wv,w} + \eta_{wv}\frac{\partial ln|\varsigma|}{\partial\phi_w} \\ &= \eta_{wv,w} + \eta_{wv}\frac{\partial ln|\varsigma|}{\partial\varsigma_{rs}}\varsigma_{rs,w} \end{aligned} \tag{3.33}$$

Using the expression for the derivative of a logarithm of a determinant

$$\begin{aligned} 0 &= \eta_{wv,w} + \eta_{wv}\eta_{sr}\varsigma_{rs,w} \\ &= \eta_{wv,w} - \eta_{wv}\varsigma_{rs}\eta_{sr,w} \end{aligned} \tag{3.34}$$

Applying eq. (2.122) we obtain

$$\begin{aligned} 0 &= \eta_{wv,w} - \eta_{sv,w}\eta_{wr}\varsigma_{rs} + c_{rvw}\eta_{sw}\varsigma_{rs} \\ &= \eta_{wv,w} - \eta_{sv,s} + c_{rvr} \end{aligned} \tag{3.35}$$

which is an identity since we have assumed that the c_{vws} are completely antisymmetric.

appendix 3.B An internal product for the Hilbert space $\mathcal{H}_2$

In order to construct an inner product that is suitable for the treatment in $\mathcal{H}_2$ (without further extensions) we have to modify (3.18) so that it remains finite and agrees for physical states with the internal product in $\mathcal{H}_1$. Such an extension is straightforward in the laboratory frame

$$< \Psi_1'|\Psi_2' > = \int \Pi_i \Pi_v d\mu(q_i')d\phi_v \Psi_1^*(q_i')\Psi_2(q_i')\delta(\phi_v) \tag{3.36}$$

where delta-functions have been introduced. More generally, we may write

$$\begin{aligned}
<\Psi'_1|\Psi'_2> &= \int \Pi_v d\mu(q'_i) d\phi_v \Psi_1^*(q'_i)\Psi_2(q'_i)\delta(G'_v) det[\frac{\partial G'_s}{\partial \phi_w}] \\
&= \int \Pi_v d\mu(q'_i) d\phi_v \Psi_1^*(q'_i)\Psi_2(q'_i)\delta(G'_v) det[\{G'_s, P'_w\}] \qquad (3.37) \\
&= \int \Pi_v d\mu(q'_i) d\phi_v \Psi_1^*(q'_i)\Psi_2(q'_i)\ |\varsigma|\delta(G'_v) det[\{G'_s, \eta_{tw} P'_t\}]
\end{aligned}$$

where $G'_s(\phi, q'_i)$ are the laboratory versions of gauge fixing functions. The transformation to the moving frame yields the inner product in the enlarged Hilbert space $\mathcal{H}_2$

$$<\Psi_1|\Psi_2>= \int \Pi_v d\mu(q_i) d\phi_v |\varsigma| \Psi_1^*(q_i, \phi_v)\Psi_2(q_i, \phi_v)\delta(G_v) det[\{G_s, f_w\}] \quad (3.38)$$

The gauge fixing functions G_v have been introduced in order to generalize the inner product. Their arbitrariness is due to the fact that it is irrelevant what happens outside the physical space. Inside the physical space, (3.37) coincides with (3.36). However, it is easy to check that the inner product (3.37) and its transformed version (3.38) connect the physical with the unphysical subspaces. Hence, although they yield correct results between physical functions, we must *impose* that we remain within the physical subspace. This can be done by a projector onto that subspace. We will not here pursue this line further, but defer this construction to section 5 from the path integral point of view in the "Coulomb gauge".

An interesting point is that in (3.37) and in (3.38) we can substitute the delta functions by:

$$\begin{aligned}
\delta(G'_v) &\to \frac{1}{\sqrt{2\pi A}} \exp[\frac{-G'^2_v}{2A}] \\
\delta(G_v) &\to \frac{1}{\sqrt{2\pi A}} \exp[\frac{-G^2_v}{2A}] \qquad (3.39)
\end{aligned}$$

This yields the correct results for physical functions provided

$$\frac{1}{\sqrt{2\pi A}} \int \Pi_v d\phi_v\ \exp[\frac{-G^{2'}_v}{2A}]\ det[\frac{\partial G'_s}{\partial \phi_w}] = 1 \qquad (3.40)$$

This equality can fail because a dependency on the physical variables q'_i may arise through the limits of integration.

This last relaxation of the gauge conditions is connected with t'Hooft's trick which we use in section 5.

4 The BRST treatment

In this section we apply the algebraic procedure developed by Batalin, Fradkin, and Vilkovisky [11] to the problem of collective coordinates. This procedure relies on the canonical formulation of the Becchi-Rouet-Stora-Tyutin transformation [10]. The BRST treatment of gauge symmetries is adequate for dealing with a system in which collective parameters and Lagrange multipliers are considered as dynamical variables on the same footing as the original coordinates. This is the case, for instance, with the classical hamiltonians (2.43) and (2.70). In order to be self-contained, we present all the relevant proofs.

In the present section we skip the classical BRST treatment of collective coordinates. It is nevertheless a simple matter to perform this treatment, which should start considering the phase-space $\mathcal{F}_3$ that is discussed in subsection 2.3.

Here we present directly a quantum mechanical derivation. Thanks to the reversed metric which we associated with the "Lagrange multipliers" Ω_v, the equivalences are proved rigorously and without the divergences and regularization problems that would be present had we introduced the usual metric at the level of these variables [23]. The pseudo-Hilbert space that results from our construction is in the representation known as "pseudo-Fock" [24], which is at the basis of the Kugo-Ojima quartet mechanism [25].

The BRST treatment introduces the ghosts from the outset. These are fermions carrying zero spin and have no direct physical meaning. On the other hand, they are very useful mathematical artifacts which are employed, for example, in the exponentiation of a determinant (cf. section 5). The introduction of ghosts does not simplify the problem per se. It does so in gauge systems because there is a supersymmetry in the enlarged Hilbert space which does not exist in the smaller space. In a similar way a person confined to a two-dimensional world could simplify the description of a problem by postulating a third dimension and realizing that the motion takes place on the surface of a sphere of constant radius.

The BRST treatment adapts to the case of very general transformations, including those which lead to constraints whose structure "constants" c_{vws} are operators, so that they do not form a true algebra (in these cases the transformations do not form a true Lie group). However we assume in this section that the c_{vws} are constants.

4.1 The general formulation

We start by considering the Hilbert space $\mathcal{H}_3$, in which physical functions satisfy the constraints (3.5) or (3.6) and (3.12). The constraints satisfy the commutation relations (3.8). We now further enlarge the Hilbert space by defining $2k$ additional odd Grassmann variables $\eta_v, \bar{\pi}_v$ (ghosts). The corresponding conjugate operators acting in the enlarged Hilbert space $\mathcal{H}_4$ are

$$\hat{\pi}_v|\Psi> = \frac{\partial}{\partial \eta_v}|\Psi>; \quad \hat{\eta}_v|\Psi> = \frac{\partial}{\partial \pi_v}|\Psi> \tag{4.1}$$

$$[\hat{\pi}_v, \hat{\eta}_w]_+ = [\hat{\bar{\pi}}_v, \hat{\bar{\eta}}_w]_+ = \delta_{vw} \tag{4.2}$$

all other anticommutators being zero.

We construct first the measure for one η and one π ghosts. We use the Berezin's integrals [26]

$$< h|g >_{gh} = i \int h^* g \, d\eta \, d\pi \tag{4.3}$$

Integration in the ghost subspace is based on the the non-vanishing overlaps

$$\int \eta \, d\eta = \int \pi \, d\pi = \int \pi\eta \, d\eta \, d\pi = 1 \tag{4.4}$$

The measure for several ghost pairs is obtained in the product form. With this definition, and using the fact that $\hat{\eta}, \hat{\pi}, \hat{\bar{\pi}}$ and $\hat{\bar{\eta}}$ are real and that the conjugation reverses the order of the Grassmann variables, we find that the ghost operators (4.1) are hermitean.

We hence have in $\mathcal{H}_4$ the two inner products $< \; | \; >^{(F)}$ and $< \; | \; >^{(f)}$ which are obtained multiplying the measures in $\mathcal{H}_3$ $< \; | \; >^{(F)}_{\mathcal{H}_3}$ (cf. eq. (3.29)) and $< \; | \; >^{(f)}_{\mathcal{H}_3}$ (cf. eq. (3.30)) by the ghost measure (4.3).

We now define the (hermitian) BRST charge operators[15]

$$\begin{aligned} \hat{Q}^{(F)} &= \hat{P}_v \hat{\pi}_v - \hat{F}_v \hat{\eta}_v \\ \hat{Q}^{(f)} &= \hat{P}_v \hat{\pi}_v - \hat{f}_v \hat{\eta}_v + \frac{i}{2} c_{vws} \hat{\eta}_v \hat{\eta}_w \hat{\pi}_s \end{aligned} \tag{4.5}$$

$\hat{Q}^{(F)}$ and $\hat{Q}^{(f)}$ are hermitean with respect to the $< \; | \; >^{(F)}$ and the $< \; | \; >^{(f)}$ metrics, respectively.

We also define the antihermitean ghost number operator

$$\hat{N}_{gh} = -\hat{N}^{\dagger}_{gh} = \hat{\eta}_v \hat{\pi}_v - \hat{\bar{\eta}}_v \hat{\bar{\pi}}_v \tag{4.6}$$

Within $\mathcal{H}_4$, the two charges can be transformed into each other by an invertible transformation (see appendix 4.A). The considerations in the remaining part of this subsection are common to both of these charges. Thus we drop the superscripts F and f.

[15]By calling $\hat{Q}$ a charge we adhere to the field-theoretical nomenclature.

The crucial properties of $\hat{Q}$ are its nilpotency[16]

$$\hat{Q}^2 = 0 \tag{4.7}$$

and its hermiticity

$$\hat{Q}^\dagger = \hat{Q} \tag{4.8}$$

We define the physical states $|ph>$ in $\mathcal{H}_4$ as those states which are scalars with respect to the transformation generated by $\hat{Q}$

$$\hat{Q}|ph>= 0 \tag{4.9}$$

and the unphysical states $|u>$

$$\hat{Q}|u>= |\chi> \neq 0 \tag{4.10}$$

We have to show that this definition of physicality, although different from the previous ones, leads to the same results.

Because of the nilpotency of $\hat{Q}$ the states $|\chi>$ so constructed are physical in the sense of eq.(4.9) but have zero norm

$$\begin{aligned} \hat{Q}|\chi> &= 0 \\ <\chi_1|\chi_2> &= <u_1|\hat{Q}^2|u_2>= 0 \end{aligned} \tag{4.11}$$

Similarly, we will define as physical those operators that commute with the charge $\hat{Q}$

$$[\hat{O}_{ph}, \hat{Q}] = 0 \tag{4.12}$$

and as unphysical the operators $\hat{O}_u$ that do not

$$[\hat{O}_u, \hat{Q}] = \hat{O}_\chi \neq 0 \tag{4.13}$$

The $\hat{O}_\chi$ operators so constructed are also physical in the sense of eq. (4.12) (as follows from Jacobi's identity). They map physical states $|ph>$ into the zero-norm states $|\chi>$

$$\hat{O}_\chi|ph>= |\chi> \tag{4.14}$$

We identify equivalence classes of physical states and operators by

$$\begin{aligned} |ph> &\rightarrow |ph> + |\chi> \\ \hat{O}_{ph} &\rightarrow \hat{O}_{ph} + \hat{O}_\chi, \end{aligned} \tag{4.15}$$

where the classes are spanned by all the different choices of $|\chi>$ and of $\hat{O}_\chi$. The same overlaps are obtained using any member of a class of states or operators. Thus the validity of the BRST formalism becomes plausible. In subsection 4.2

[16]This statement can be directly verified from the definitions (4.5). It is also demonstrated in appendix 4.A.

we show more rigorously that within each class of states and operators there is one member for which the equivalence between the BRST- and the treatment in the original Hilbert space with the original operators is straightforwardly verified. Hence, from the discussion in this section, the results are valid for any member of the equivalence classes.

In particular, we can use the effective hamiltonian

$$\hat{H}_{BRST} = \hat{H} + [\hat{\rho}, \hat{Q}]_+ \tag{4.16}$$

where $\hat{\rho}$ is an arbitrary hermitian odd-Grassmann operator. The hamiltonian H_{BRST} is hermitean and, from the preceding discussion, yields the same results as the original hamiltonian in $\mathcal{H}_1$. A possible choice[17] for $\hat{\rho}$ is motivated by the analogy with the covariant gauge in Yang-Mills theory

$$\hat{\rho} = \hat{\Omega}_v \hat{\bar{\pi}}_v + \hat{\bar{\eta}}_v \left(\frac{1}{B_v}\hat{G}_v - \frac{A_v}{2B_v^2}\hat{P}_v\right) \tag{4.17}$$

Using $\hat{Q}^{(f)}$,the effective hamiltonian becomes

$$\begin{aligned} \hat{H}_{BRST} \quad &= \quad \hat{H} + i\hat{\bar{\pi}}_v \hat{\pi}_v - \hat{\Omega}_v \hat{f}_v + \frac{1}{B_v}\hat{G}_v \hat{P}_v - \frac{A_v}{2B_v^2}\hat{P}_v^2 \\ &\quad - \frac{1}{B_v}\hat{\bar{\eta}}_v [\hat{G}_v, \hat{f}_w] \hat{\eta}_w + ic_{vws}\hat{\Omega}_s \hat{\bar{\pi}}_v \hat{\eta}_w \end{aligned} \tag{4.18}$$

The A_v, B_v are arbitrary constants and should not appear in any physical final result[18].

The $\hat{G}_v$ are functions that are chosen such that $[\hat{G}_v, \hat{J}_w]$ may be approximated by $i\delta_{vw}$ plus small terms with vanishing expectation value. Their meaning may be clarified by performing a minimization of the hamiltonian H_{BRST} , which requires the vanishing of the expectation values

$$\begin{aligned} < [\hat{H}_{BRST}, \hat{P}_v] > \quad &= \quad 0 \\ < [\hat{H}_{BRST}, \hat{\Omega}_v] > \quad &= \quad 0 \\ < [\hat{H}_{BRST}, \hat{f}_v] > \quad &= \quad 0 \end{aligned} \tag{4.19}$$

At this stage we replace the expectation value of the product of two operators by the product of the corresponding expectation values. Within this semiclassical approximation, eqs. (4.19) yield

[17] A slightly modified form is used if there is a partial breakdown of symmetry (cf. subsection 8.2).

[18] In field theory the value of some of these constants is fixed by requirements of covariance which do not exist in our case.

$$
\begin{aligned}
<\hat{J}_v> &= <\hat{I}_v> \\
\frac{A_v}{B_v} <\hat{P}_v> &= <\hat{G}_v> \\
c_{vws} <\hat{f}_v><\hat{\Omega}_s> &= \frac{1}{B_v} <\hat{P}_v>
\end{aligned}
\tag{4.20}
$$

Combining these equations we obtain

$$
<G_v>= 0 \tag{4.21}
$$

Usually the G_v are functions of the original coordinates of the system, and this last equation fixes the position of the system relative to the moving frame of reference. Therefore the G_v play the role of the gauge fixing functions (2.78)[10].

We operate with this hamiltonian in the usual way (from the previous discussion it yields the same physical results as the original one. To calculate the overlaps of physical operators which are originally given in the laboratory frame, we take the frame using the transformation inverse to (3.2). If the operators are not invariant with respect to the transformation group, they acquire a dependance on the collective variables, and hence also act on the collective part of the wavefunctions.

Although many of the developments associated with the quantification of gauge theories were originally inspired by functional methods, in this section we have deliberately avoided using them. Nevertheless, at least for simple cases of collective coordinates associated with ordinary Lie groups, we may construct expressions such as (4.18) within the Faddeev-Popov formalism of section 5, although loosing much of the algebraic insight and generality.

For the time being the reader may feel satisfied with the plausibility arguments of this subsection. He may proceed directly to clarify the use of the BRST treatment through the illustrative examples presented in sections 7 and 8.

We derive now a simple argument which is frequently used in the BRST formalism. Let $\hat{A}_X$ be a "null" operator (i.e. $\hat{A}_X = [\hat{Q}, \hat{O}_u]$, eq. (4.13)) which, in addition, commutes with $\hat{H}_{BRST}$. Thus physical wavefunctions may be labelled by the eigenvalues of $\hat{A}_X$ and $\hat{H}_{BRST}$

$$
\begin{aligned}
\hat{A}_X|\epsilon,\alpha> &= \alpha|\epsilon,\alpha> \\
\hat{H}_{BRST}|\epsilon,\alpha> &= \epsilon|\epsilon,\alpha> \\
\hat{Q}|\epsilon,\alpha> &= 0
\end{aligned}
\tag{4.22}
$$

If $\alpha \neq 0$, then

$$
|\epsilon,\alpha> = \frac{1}{\alpha}\hat{A}_X|\epsilon,\alpha>= \frac{1}{\alpha}\hat{Q}\hat{O}_u|\epsilon,\alpha>
$$

[10] In the case of three dimensional rotations we may take $\hat{G}_v = \hat{Q}_{ws}$ (v,w,s cyclical), as in eq. (2.105). Thus the condition (4.21) fixes the moving frame to principal axis.

$$= \hat{Q}[\frac{1}{\alpha}\hat{O}_{\chi}|\epsilon,\alpha>] \tag{4.23}$$

hence if $\alpha \neq 0$, $|\epsilon,\alpha>$ is a "null" state (4.13). Therefore, only states annihilated by $\hat{A}_{\chi}$ can have non-zero norm (and are thus the only ones of physical interest).

4.2 The equivalence between the BRST treatment in $\mathcal{H}_4$ and the usual treatment in $\mathcal{H}_1$

The proof of the equivalence of the BRST treatment in $\mathcal{H}_4$ with the treatment in the original space $\mathcal{H}_1$ is divided in two parts : we first use some results of appendix 4.A to show that it suffices to consider the case of constraints $\hat{F}'_v$ in involution (with the associated metric $< \mid >^{(F)}$) and in the laboratory frame in which these constraints take the simple form (3.14). Secondly, we prove the equivalence for such a case.

4.2.1 The abelianization of constraints

In appendix 4.A [20] we construct explicitly an operator $\hat{\hat{R}}$ which transforms the two charges (4.5) into each other

$$\hat{\hat{R}}\hat{Q}^{(F)}\hat{\hat{R}}^{-1} = \hat{Q}^{(f)} \tag{4.24}$$

This operator satisfies

$$[\hat{\hat{R}}, \hat{N}_{gh}] = 0 \tag{4.25}$$

so that $\hat{\hat{R}}$ preserves the ghost number. It takes the same form in the laboratory and rotated frames since we have that

$$[\hat{\hat{R}}, \hat{T}_{\phi}] = 0 \tag{4.26}$$

and hence

$$\hat{\hat{R}}\hat{Q}'^{(F)}\hat{\hat{R}}^{-1} = \hat{Q}'^{(f)} \tag{4.27}$$

If $\hat{O}'_{ph}$ is a physical operator that acts within $\mathcal{H}_1$ and acts only on the original variables (e.g. the hamiltonian) and $\hat{O}_{ph}$ is its transformed version, we have

$$[\hat{\hat{R}}, \hat{O}'_{ph}] = [\hat{\hat{R}}, \hat{O}_{ph}] = 0 \tag{4.28}$$

Given two physical states $|\psi_i^{(f)}>$ $(i = 1,2)$ we construct their laboratory versions

$$|\psi_i^{(f)'}> = \hat{T}^{-1}|\psi_i^{(f)}> \tag{4.29}$$

---

[20] See also ref. [27]

We also construct the states (cf. eq. (4.68))

$$|\psi_i^{(F)'}>=\hat{\hat{R}}^{-1}|\psi_i^{(f)'}>=T^{-1}\hat{\hat{R}}^{-1}|\psi_i^{(f)}> \tag{4.30}$$

It is easy to see that

$$\hat{Q}^{(f)}|\psi_i^{(f)}>=0\Leftrightarrow\hat{Q}^{(f)'}|\psi_i^{(f)'}>=0\Leftrightarrow\hat{Q}^{(F)'}|\psi_i^{(F)'}>=0 \tag{4.31}$$

For operators satisfying eq. (4.28) (like the hamiltonian) we have (cf. (4.73))

$$<\psi_1^{(f)}|\hat{O}_{ph}|\psi_2^{(f)}>^{(f)}=<\psi_1^{(f)'}|\hat{O}_{ph}'|\psi_2^{(f)'}>^{(f)}=<\psi_1^{(F)'}|\hat{O}_{ph}'|\psi_2^{(F)'}>^{(F)} \tag{4.32}$$

Hence, we have shown that we can use the transformations T and $\hat{\hat{R}}$ to map the problem into a problem in the laboratory frame with commuting constraints and their associated metric $<\ |\ >^{(F)}$. This completes the first part of the proof.

4.2.2 The proof for commuting constraints

We work in the laboratory frame. Since in this frame the original Hilbert space $\mathcal{H}_1$ is automatically factorized, we consider only the (spurious) space associated with the collective variables, the Lagrange multipliers and the ghosts. Moreover, we work with the set of constraints

$$\hat{F}_v'=-\hat{P}_v'\ ;\ \hat{P}_v \tag{4.33}$$

which are in involution (cf. eqs. (3.8)). Consequently, we may restrict ourselves to the case of a single pair of constraints $\hat{F}'$ and $\hat{P}$, the generalization to several constraints in involution being trivial.

It will be advantageous to use the following linear combinations of the ghost operators

$$\begin{aligned} a &= \frac{1}{\sqrt{2}}(\hat{\bar{\pi}}-i\hat{\eta}) \qquad b=\frac{1}{\sqrt{2}}(\hat{\pi}+i\hat{\bar{\eta}}) \\ \bar{a} &= \frac{1}{\sqrt{2}}(i\hat{\pi}+\hat{\bar{\eta}}) \qquad \bar{b}=\frac{1}{\sqrt{2}}(-i\hat{\bar{\pi}}+\hat{\eta}) \end{aligned} \tag{4.34}$$

which satisfy the following non-vanishing anticommutation relations

$$[a,\bar{a}]_+=[b,\bar{b}]_+=1 \tag{4.35}$$

The ghost number operator (4.6) is written

$$\hat{N}_{gh}=\bar{b}b-\bar{a}a \tag{4.36}$$

We define the (norm one) ghost vacuum φ_o^{gh} as :

$$\varphi_o^{gh}\equiv\frac{1}{\sqrt{2}}(\bar{\pi}-i\eta) \qquad a\varphi_o^{gh}=b\varphi_o^{gh}=0 \tag{4.37}$$

and we create ghost states[21] by applying $\bar{a}, \bar{b}$ to φ_o^{gh}.

We now turn our attention to the spurious phonon sector and define the creation and destruction operators

$$\Gamma_\phi \equiv \frac{1}{\sqrt{2}}(\hat{\phi} + i\hat{P}') \;\; ; \;\; \Gamma_\phi^\dagger \equiv \frac{1}{\sqrt{2}}(\hat{\phi} - i\hat{P}') \;\; ; \;\; [\Gamma_\phi, \Gamma_\phi^\dagger] = 1 \tag{4.38}$$

$$\Gamma_\Omega^\dagger \equiv \frac{1}{\sqrt{2}}(\hat{\Omega} - i\hat{P}) \;\; ; \;\; \Gamma_\Omega \equiv \frac{1}{\sqrt{2}}(\hat{\Omega} + i\hat{P}) \;\; ; \;\; [\Gamma_\Omega, \Gamma_\Omega^\dagger] = 1 \tag{4.39}$$

We use the considerations made in subsection 3.2 in order to define a basis for the degrees of freedom corresponding to the collective coordinates and Lagrange multipliers. Therefore, we complete the construction of a basis for $\mathcal{H}_4$ by requiring the vacuum of the spurious phonons to satisfy

$$\Gamma_\phi |0_\phi 0_\Omega > = \Gamma_\Omega |0_\phi 0_\Omega > = 0 \tag{4.40}$$

and the state having m_ϕ and n_Ω bosons as:

$$|m_\phi \, n_\Omega > \equiv \frac{1}{\sqrt{n_\Omega ! m_\phi !}} [\Gamma_\phi^\dagger]^{m_\phi} [\Gamma_\Omega]^{n_\Omega} |0_\phi 0_\Omega > \tag{4.41}$$

Notice that

$$< m'_\phi \, n'_\Omega | m_\phi \, n_\Omega > = (-1)^{n_\Omega} \delta_{n_\Omega n'_\Omega} \delta_{m_\phi m'_\phi} \tag{4.42}$$

These equations have the effect of defining a (pseudo) Hilbert space as in the previous section.

A basis for $\mathcal{H}_4$ can now be constructed as the product of wavefunctions of the physical variables times ghost states times states of the form (4.41). However it is more convenient to use an alternative basis, which is unitarily related to the previous one

$$\Gamma_o^\dagger = \frac{1}{\sqrt{2}}(\Gamma_\Omega + i\Gamma_\phi) \;\; ; \;\; \Gamma_1^\dagger = \frac{i}{\sqrt{2}}(\Gamma_\phi^\dagger - i\Gamma_\Omega^\dagger)$$

$$\Gamma_o = \frac{1}{\sqrt{2}}(\Gamma_\Omega^\dagger - i\Gamma_\phi^\dagger) \;\; ; \;\; \Gamma_1 = -\frac{i}{\sqrt{2}}(\Gamma_\phi + i\Gamma_\Omega) \tag{4.43}$$

Clearly (note the commutation of Γ_o)

$$[\Gamma_o^\dagger, \Gamma_o] = [\Gamma_1, \Gamma_1^\dagger] = 1 \tag{4.44}$$

---

[21] The vacuum ghost state and the state $\bar{a}\bar{b}\varphi_o^{gh}$ have zero ghosts; the state $\bar{a}\varphi_o^{gh}$ carries ghost number -1 and the state $\bar{b}\varphi_o^{gh}$, ghost number 1. Since $a^\dagger = i\bar{b} \neq \bar{a}$ and $b^\dagger = -i\bar{a} \neq \bar{b}$ the states created by $\bar{a}$ and $\bar{b}$ do not have a positively defined norm. If $|n_1>$ $|n_2>$ are states with n_1 and n_2 ghosts, respectively, then

$$< n_1 | n_2 > \;\; (n_1 + n_2) = 0$$

In particular, only zero-ghost states can have non-zero norm.

The operators Γ_1, Γ_o annihilate the new vacuum

$$\Gamma_1|0_1 0_o> = \Gamma_o|0_1 0_o> = 0 \tag{4.45}$$

and the state having m_1 and n_o bosons is given by

$$|m_1\ n_o> \equiv \frac{1}{\sqrt{n_o! m_1!}}[\Gamma_1^\dagger]^{m_1}[\Gamma_o^\dagger]^{n_o}|0_1 0_o> = S|m_\phi\ n_l> \tag{4.46}$$

It is easy to construct explicitly the unitary transformation S between these states and those in eq. (4.41).

The pseudo Hilbert space $\mathcal{H}_4$ is generated by states of the form

$$|m_1\ n_o\ \alpha\ \beta\ r> = |\Psi_r^D(q_i')> \otimes |m_1\ n_o> \otimes \bar{a}^\alpha \bar{b}^\beta \varphi_o^{gh} \tag{4.47}$$

where the $\Psi_r^D(q_i')$ constitute a basis of wavefunctions of the physical space.

In terms of the new operators (4.34) and (4.43) the BRST charge reads

$$\hat{Q}'^{(F)} == \frac{i}{\sqrt{2}}a(\Gamma_o^\dagger + \Gamma_1^\dagger) + \frac{1}{\sqrt{2}}\bar{b}(\Gamma_o + \Gamma_1) \tag{4.48}$$

We also define the scaling ("null") operator [24]

$$\begin{aligned} \hat{D} &= \hat{D}^\dagger = -\hat{P}\hat{\phi} - \hat{F}'\hat{\Omega} - i(\hat{\bar{\pi}}\hat{\pi} + \hat{\bar{\eta}}\hat{\eta}) = [\hat{Q}'^{(F)}, -\hat{\phi}\hat{\bar{\eta}} + \hat{\Omega}\hat{\bar{\pi}}]_+ \\ &= \Gamma_1^\dagger\Gamma_1 - \Gamma_o^\dagger\Gamma_o + \bar{a}a + \bar{b}b \end{aligned} \tag{4.49}$$

Since $\hat{Q}'^{(F)}, \hat{N}_{gh}$ and $\hat{D}$ commute, physical states can be classified simultaneously by their ghost number and the eigenvalue of $\hat{D}$. We can now apply the result obtained at the end of subsection 4.1: only physical states annihilated by $\hat{D}$ do not have zero norm [24] (i.e., they are not the result of applying $\hat{Q}'^{(F)}$ to some state in $\mathcal{H}_4$.). Since only states of the form (4.47) with $n = m = \alpha = \beta = 0$ satisfy this requirement, *there is a single representative of this form for every physical state, and it has well defined norm.* Any physical state can be taken by addition of zero norm states to have only components of this form.

Note the role played by the reversed metric at bosonic level: had we considered states created by Γ_o and destroyed by $\Gamma_o^\dagger$ (which would lead to having the ordinary metric at the level of the bosons) we would have positive and negative eigenvalues for $\hat{D}$, and thus an infinity of states annihilated by $\hat{D}$. The situation with the metric (4.44) is entirely similar to the case of electromagnetism in which one introduces the temporal and longitudinal bosons (although we do not have covariance arguments to follow here).

Given any operator that can be written in terms of creation and destruction operators (in normal order)

$$\hat{O} = \sum O_{r\hat{r}s\hat{s}\gamma\hat{\gamma}\delta\hat{\delta}}(q_i', \frac{\partial}{\partial q_i'})\ \Gamma_o^{\dagger r}\Gamma_1^{\dagger s}\Gamma_o^{\hat{r}}\Gamma_1^{\hat{m}}\bar{a}^\gamma\bar{b}^\delta a^{\hat{\gamma}} b^{\hat{\delta}} \tag{4.50}$$

Clearly the only relevant term as far as overlaps are concerned between non-zero norm physical states ($n = m = \alpha = \beta = 0$) is the one that has $r = s = \gamma = \delta = 0$ and $\tilde{r} = \tilde{s} = \tilde{\gamma} = \tilde{\delta} = 0$, which we denote by $\hat{O}^D$. The other terms map physical into zero norm states (or zero). It is easy to see that the overlaps of $\hat{O}^D$ between two different states of the form (4.47) is reduced to $< \Psi^D_{r'} | \hat{O}^D | \Psi^D_r >$, which is well defined.

In particular, the operator can be the hamiltonian. Thus we have found a correspondence between the eigenstates of H_{BRST} in $\mathcal{H}_4$, the Dirac eigenstates in $\mathcal{H}_3$ and the original eigenstates in $\mathcal{H}_1$.

All the previous manipulations (using the constraints in involution, working in the laboratory frame, obtaining the standard form of the wavefunctions) were carried out for demonstration purposes. The present demonstration guarantees that we never need to perform them in practice. In the applications we always use the constraints $\hat{f}_v$ *and thus we omit the superscript* (f).

appendix 4.A The transformation properties of the charge $\hat{Q}$

In this appendix we construct explicitly an operator $\hat{\hat{R}}$ such that

$$\hat{\hat{R}}\hat{Q}'^{(F)}\hat{\hat{R}}^{-1} = \hat{Q}'^{(f)} \quad (4.51)$$

The existence of such operator is a well known fact [28,13,27]. This does not imply that the operator $\hat{\hat{R}}$ transforms the constraints $\hat{F}_v$ into the $\hat{f}_v$. This cannot be the case, since the operators $\hat{F}_v$ commute among themselves and the $\hat{f}_v$ do not.

We proceed in two steps: first we construct a unitary operator $\hat{R}$ such that

$$\hat{R}\hat{Q}'^{(F)}\hat{R}^{\dagger} = \hat{P}_v\hat{\pi}_v + \frac{1}{2}[\eta_{wv}, \hat{p}'_w]_+\hat{\eta}_v + \frac{i}{2}c_{vws}\hat{\eta}_v\hat{\eta}_w\hat{\pi}_s \quad (4.52)$$

The r.h.s. is "almost" $\hat{Q}'^{(f)}$ (in the laboratory frame), but for the fact that $\hat{p}_v$ does not commute with η_{vw}. The unitary transformation $\hat{R}$ maps $\hat{Q}'^{(F)}$ into an operator which is hermitian with respect to the norm $< \mid >^{(F)}$ associated with the F_v constraints. Clearly, $\hat{R}$ is the part of the transformation which makes contact with a classical canonical transformation.

Let us put

$$\hat{R} = exp[A_{rs}\hat{j}_{rs}] \quad (4.53)$$

where the operator $\hat{j}_{vw}$ is

$$\hat{j}_{rs} = -\hat{j}^{\dagger}_{rs} \equiv \frac{1}{2}(\hat{\eta}_r\hat{\pi}_s - \hat{\pi}_s\hat{\eta}_r) \quad (4.54)$$

The matrix $A(\phi_v)$ is defined such that

$$\begin{aligned}(exp[A])_{wv} &= \eta_{vw} \\ (exp[-A])_{wv} &= \varsigma_{vw}\end{aligned} \quad (4.55)$$

which is always possible if $|\varsigma| \neq 0$. The $\hat{j}_{rs}$ form a U(k) algebra and hence $\hat{R}$ belongs to U(k). From the commutation properties of the $\hat{j}_{rs}$ it is straightforward to see that

$$\begin{aligned}\hat{R}\hat{\eta}_v\hat{R}^{\dagger} &= \eta_{vw}\hat{\eta}_w \\ \hat{R}\hat{\pi}_v\hat{R}^{\dagger} &= \varsigma_{vw}\hat{\eta}_w\end{aligned} \quad (4.56)$$

The only other operators that are affected by this transformation are the $\hat{p}'_v$.

$$\hat{R}\hat{p}'_v\hat{R}^{\dagger} == -i\hat{R}(\frac{\partial\hat{R}^{\dagger}}{\partial\phi_v}) + \hat{p}'_v = -i\xi_{rsv}(\phi_t)\hat{j}_{rs} + \hat{p}'_v \quad (4.57)$$

which is of the form (3.5) (the ξ_{rsv} playing a role analogous to the ς_{wv}) but for the case of group generated by the $\hat{j}_{rs}$. In order to calculate the ξ_{rsv} we differentiate eq. (4.56) and we obtain

$$\eta_{ws}[\hat{\eta}_s, \hat{R}(\frac{\partial R^\dagger}{\partial \phi_v})] = \eta_{ws.v}\hat{\eta}_s \tag{4.58}$$

Hence using (4.57)

$$\xi_{rsv} = -\varsigma_{sw}\eta_{wr.v} \tag{4.59}$$

so that

$$\hat{R}\hat{P}'_v\hat{R}^\dagger = \hat{P}'_v + i\varsigma_{sw}\eta_{wr.v}\hat{j}_{rs} \tag{4.60}$$

Finally, for the BRST charge we obtain

$$\begin{aligned}
\hat{R}\hat{Q}'^{(F)}\hat{R}^\dagger &= P_v\hat{\bar{\pi}}_v + \frac{1}{2}\hat{R}[\hat{P}'_v, \hat{\eta}_v]_+\hat{R}^\dagger \\
&= P_v\hat{\bar{\pi}}_v + \frac{1}{2}[\hat{P}'_v, \eta_{vw}]\hat{\eta}_w + \frac{i}{2}\varsigma_{sw}\eta_{wr.v}\eta_{vt}[\hat{j}_{rs}, \hat{\eta}_t]_+] \\
&= P_v\hat{\bar{\pi}}_v + \frac{1}{2}[\hat{P}'_v, \eta_{vw}]\hat{\eta}_w + \frac{i}{2}\varsigma_{sw}(\eta_{wr.v}\eta_{vt} - \eta_{wt.v}\eta_{vr})[\hat{j}_{rs}, \hat{\eta}_t]_+ \\
&= P_v\hat{\bar{\pi}}_v + \frac{1}{2}[\hat{P}'_v, \eta_{vw}]\hat{\eta}_w - \frac{i}{4}c_{rst}[\hat{j}_{rt}, \hat{\eta}_s]_+ \\
&= \hat{P}_v\hat{\bar{\pi}}_v + \frac{1}{2}[\eta_{wv}, \hat{P}'_w]\hat{\eta}_v + \frac{i}{2}c_{vws}\hat{\eta}_v\hat{\eta}_w\hat{\bar{\pi}}_s
\end{aligned} \tag{4.61}$$

where we have used the fact that $[\hat{j}_{rs}, \hat{\eta}_t]_+ = -[\hat{j}_{ts}, \hat{\eta}_r]_+$ and eq. (2.121).

The next step is to bring the charge to the form that it finally takes in the space where the $\hat{f}_v$ are hermitian. It is easy to see that this additional transformation should be non-unitary (and it is essentially quantum mechanical). Let us calculate

$$\begin{gathered}
-\tfrac{1}{2}|\varsigma|^{-1/2}[\eta_{vw}\hat{P}'_v]_+|\varsigma|^{1/2} = \\
\eta_{vw}\hat{P}'_v - \tfrac{i}{2}(\tfrac{\partial \ln|\varsigma|}{\partial \phi_v} + \eta_{vw.v}) = \eta_{vw}\hat{P}'_v
\end{gathered} \tag{4.62}$$

The last term vanishes by virtue of (3.33). From (4.62) we have

$$\hat{\hat{R}}\hat{Q}'^{(F)}\hat{\hat{R}}^{-1} = \hat{Q}'^{(f)} \tag{4.63}$$

with

$$\hat{\hat{R}} \equiv \hat{R}|\varsigma|^{-1/2} \tag{4.64}$$

Finally, it is trivial to check that

$$[\hat{\hat{R}}, \hat{T}(\phi_v)] = 0 \tag{4.65}$$

From this equation we immediately see that the transformation is also valid in the rotating frame

$$\hat{\hat{R}}\hat{Q}^{(F)}\hat{\hat{R}}^{-1} = \hat{Q}^{(f)} \tag{4.66}$$

The transformation $\hat{\hat{R}}$ also preserves the ghost number

$$[\hat{\hat{R}}, \hat{N}_{gh}] = 0 \tag{4.67}$$

If we are given a set states $|\psi_i^{(f)}>$ in the space with the constraints f_v (and their associated metric $<\ |\ >^{(f)}$) we obtain the corresponding state in the space with the constraints F_v (and their associated metric $<\ |\ >^{(F)}$) through

$$|\psi_i^{(F)}>= \hat{\hat{R}}^{-1}|\psi_i^{(f)}> \tag{4.68}$$

$$\hat{Q}^{(F)}|\psi_i^{(F)}>= \hat{Q}^{(F)}\hat{\hat{R}}^{-1}|\psi_i^{(f)}>= \hat{\hat{R}}^{-1}\hat{Q}^{(f)}|\psi_i^{(f)}> \tag{4.69}$$

Now, using the unitarity of $\hat{R}$ and eqs. (3.19) and (3.20), we obtain

$$\begin{aligned} <\psi_1^{(F)}|\psi_2^{(F)}>^{(F)} &= <\psi_1^{(F)}|\hat{R}^\dagger\hat{R}|\psi_2^{(F)}>^{(F)} \\ &= <\psi_1^{(F)}|\hat{\hat{R}}^\dagger|_S|\hat{\hat{R}}|\psi_2^{(F)}>^{(F)} \\ &= <\psi_1^{(F)}|\hat{\hat{R}}^\dagger\hat{\hat{R}}|\psi_2^{(F)}>^{(f)} \\ &= <\psi_1^{(f)}|\psi_2^{(f)}>^{(f)} \end{aligned} \tag{4.70}$$

If $\hat{O}$ is an operator in the space with measure $<\ |\ >^{(f)}$

$$\hat{\hat{R}}\hat{O}^{(F)}\hat{\hat{R}}^{-1} = \hat{O}^{(f)} \tag{4.71}$$

and

$$[\hat{O}^{(F)}, \hat{Q}^{(F)}] = 0 \Rightarrow [\hat{O}^{(f)}, \hat{Q}^{(f)}] = 0 \tag{4.72}$$

Finally, the result (4.70) generalizes to

$$<\psi_1^{(F)}|\hat{O}^{(F)}|\psi_2^{(F)}>^{(F)}=<\psi_1^{(f)}|\hat{O}^{(f)}|\psi_2^{(f)}>^{(f)} \tag{4.73}$$

The relation

$$[\hat{O}^{(f)}, \phi_v] = 0 \tag{4.74}$$

implies

$$\hat{O}^{(F)} = \hat{O}^{(f)} \equiv \hat{O} \tag{4.75}$$

which is the case of most operators we are interested in, in particular the hamiltonian.

The technical use of these results is that we can map a problem with the constraints f_v and the inner product $< \mid >^{(f)}$ into a problem in a space with the constraints F_v and the inner products $< \mid >^{(F)}$, where proofs are much simpler.

As an example, we see that the (trivially verifiable) result

$$(\hat{Q}^{(F)})^2 = 0 \tag{4.76}$$

using the transformation $\hat{\hat{R}}$ implies the less obvious result

$$(\hat{Q}^{(f)})^2 = 0 \tag{4.77}$$

5 The quantification procedure of Faddeev-Popov

In section 2 we have presented the classical problem of collective variables as a problem of a particular type of classical gauge fields. We now return to that presentation (specially to subsection 2.3) and proceed with the quantification. An immediate procedure would be to make the correspondence between the Dirac brackets and the commutators. However, here we apply an alternative method using path-integrals [5]. Formally, it bears no obvious relation with the procedure followed in sections 3 and 4. We derive effective hamiltonians in two different gauges, which are the mechanical analogues of the Coulomb and Lorentz gauges. For this last gauge we reproduce the BRST hamiltonian (4.18).

5.1 The path-integral procedure

The quantization of a gauge field requires a formalism which takes into account its geometrical nature. Such a formalism is conveniently provided by the technique of path integrals. In the calculation of the path integrals over the extended phase-space, we must require that each physical trajectory is counted once and that only trajectories within the constraining hypersurface are included.

Let us write a path integral for the problem defined by the original lagrangian in the phase-space $\mathcal{F}_1$.

$$Z = \int \Pi_i D[p'_i] D[q'_i] \;\; exp[i \int dt (p'_j \dot{q}'_j - H(p'_i, q'_i))] \tag{5.1}$$

We insert in this amplitude the unit factor, as expressed by means of a functional integral

$$\begin{aligned} 1 &= \int \Pi_v D[P'_v] D[\phi_v] \;\; \delta(P'_v) \;\; \delta(\phi'_v) \\ &= \int \Pi_v D[P'_v] D[\phi_v] \;\; \delta(\eta_{wv} P'_w) \;\; \delta(G'_v) \;\; \Delta'_{FP} \end{aligned} \tag{5.2}$$

where Δ'_{FP} is the Jacobian corresponding to the change in the argument of both δ-functions. The matrix η has been defined before (eq. (2.61)). The functions G'_v must obviously satisfy the condition that Δ'_{FP} should not vanish.

The multiplication of the path integrals (5.1) and (5.2) yields a path integral that is equivalent to Z but in the extended phase space $\mathcal{F}_2$ of coordinates $(q'_i(t), \phi'_v(t))$ and momenta $(p'_i(t), P'_v(t))$.

The jacobian Δ'_{FP} is called the Faddeev-Popov determinant. Since the integration variables in the path integral (5.2) are labelled by the indices (v, t) ,the Faddeev-Popov determinant is of dimension $2kN$ where k is the group dimension and N the number of time intervals along the path integration. In what follows, for the sake of clarity, we may use either discreet times t_a $(a = 1, ..., N)$ or continuous times t.

The functions $G'_v(t)$ may depend on the coordinates at different times. Such is the case, for instance, if $G'_v(t)$ depends on the time derivative of the coordinates. However we assume for simplicity that it is not a function of the collective momenta

$$\frac{\partial G'_v(t_a)}{\partial \phi'_w(t_b)} \neq 0 \qquad \frac{\partial G'_v(t_a)}{\partial \mathcal{P}'_w(t_b)} \rightarrow \frac{\delta G'_v(t)}{\delta \mathcal{P}'_w(t')} = 0 \tag{5.3}$$

In consequence, the Faddeev-Popov determinant can be written in the time-indices t_a, t_b and the discrete indices v, w

$$\begin{aligned} \Delta'_{FP} &= \begin{vmatrix} \frac{\delta G'_v(t_a)}{\delta \phi'_w(t_b)} & 0 \\ \frac{\delta \eta_{sv}(t_a)}{\delta \phi'_w(t_a)} \mathcal{P}'_s(t_a)\delta_{t_b t_a} & \eta_{wv}(t_a)\delta_{t_b t_a} \end{vmatrix} \\ &= det\left[\frac{\partial G'_v(t_a)}{\partial \phi'_s(t_b)}\eta_{sw}(t_b)\right] = det[\{G'_v(t_a)\ ,\ \eta_{sw}(t_b)\mathcal{P}'_s(t_b)\}] \\ &\rightarrow det[G'_v(t), \eta_{sw}(t')\mathcal{P}'_s(t')] \end{aligned} \tag{5.4}$$

We perform now the transformation to the moving frame of reference. If the jacobian of the transformation from the laboratory to the moving frame has the value one, the volume element in both path integrals (5.1) and (5.2) remains invariant. The collective coordinate is not changed by the transformation (2.58) and the transformed derivative $\partial/\partial\phi'_v$ is the constraint operator F_v (c.f. eqs. (2.59)). Therefore,

$$-T^{-1}\ \eta_{wv}\mathcal{P}'_w = \eta_{wv}F_w = f_v \tag{5.5}$$

and thus the unit factor (5.2) is transformed into

$$\begin{aligned} 1 &= \int \Pi_v D[\mathcal{P}_v]D[\phi_v]\ \delta(f_v)\ \delta(G_v)\ \Delta_{FP} \\ &= \int \Pi_v D[I_v]D[\phi_v]\ |\varsigma|\ \delta(f_v)\ \delta(G_v)\ \Delta_{FP} \end{aligned} \tag{5.6}$$

where the matrix ς has been introduced in (2.55) and $|\varsigma|$ yields the volume element in the collective subspace[22 23].

The rotated Faddeev-Popov determinant reads

$$\begin{aligned} \Delta_{FP} &= det[\{G_v(t_a), f_w(t_b)\}] \equiv det[\Delta_{vw}(t_a, t_b)] \\ &\rightarrow det[\Delta_{vw}(t, t')] \end{aligned} \tag{5.7}$$

[22]See the footnote on p. 28.

[23]A similar factor $\delta(G_v)$ appears in the extension from the Hilbert space $\mathcal{H}_1$ to $\mathcal{H}_2$ in the treatment outlined in appendix 3.B. This factor ensures there the convergence of the integrals in $\mathcal{H}_2$. The condition $\delta(f_v)$ restraining the integral to the constraining hypersurface is equivalent to a projection over the physical subspace in the formulation of section 3.

For future use we note that the matrix $\Delta_{vw}(t_a, t_b)$ is the derivative of a gauge transformation of the gauge fixing function (c.f. eq.(2.83))

$$\begin{aligned}\Delta_{vw}(t_a, t_b) &= \frac{\partial T(\alpha_s)}{\partial \alpha_w(t_b)} G_v(t_a)|_{\alpha_s=0} \\ &= \frac{\partial}{\partial \alpha_w(t_b)} exp[-\alpha_s(t_c)\{f_s(t_c), \;\}] G_v(t_a)|_{\alpha_s=0} \\ &\rightarrow \frac{\delta T(\alpha_s)\; G_v(t)}{\delta \alpha_w(t')}|_{\alpha_s=0} \end{aligned} \qquad (5.8)$$

The kinetic terms in the lagrangian transform as in (2.68). Therefore, the path integral in the extended phase-space $\mathcal{F}_2$ finally reads

$$\begin{aligned} Z &= \int \Pi_{v,i} D[I_v] D[\phi_v] D[p_i] D[q_i]\; |\varsigma|\; \delta(f_v)\; \delta(G_v)\; \Delta_{FP} \\ & \quad exp[i \int dt(p_j \dot{q}_j + P_w \dot{\phi}_w - H)] \end{aligned} \qquad (5.9)$$

where we have made effective the constraints $f_s = 0$ in the exponent. As mentioned in section 2, all possible physical trajectories take place in the constraining hypersurface (not only the classical trajectory). This is ensured by the factor $\delta(f_v)$ in the integrand. The gauge conditions (2.30) guarantee that only one physical trajectory is selected form each class of equivalent trajectories. Therefore the functions G_v appearing in (5.9) are interpreted as the gauge fixing functions (2.78).

The strategy is now to exponentiate the Faddeev-Popov determinant, the gauge condition and the constraint, in order to obtain effective lagrangians and hamiltonians.

Because of the gauge invariance of the problem, the path integral does not change if we substitute the gauge fixing functions G_v by $G_v + \gamma_v$ where γ_v is a time-dependent constant. We can integrate over different values of γ_v with a gaussian weight,

$$\int \Pi_v D[\gamma_v]\; \delta(G_v + \gamma_v)\; exp[-i \int \frac{\gamma_w^2}{2A_w} dt] = exp[-i \int \frac{G_v^2}{2A_v} dt] \qquad (5.10)$$

By choosing a superposition of gauges centered around $G_v = 0$, these former gauge conditions have been relaxed. Since the A_v are arbitrary constants, physical results should not depend on their value (provided it is finite). This exponentiation is known as t'Hooft's trick.

The substitution of $\delta(G_v)$ by $\mathcal{N}\; exp[-\int G_v^2/2A]$, where $\mathcal{N}$ is a suitable normalization constant, could have made from the outset in eq. (5.2). However, to prove the equality between the first and second lines we must make a change of variables from (ϕ_v, P_v') to (G_v', P_v'). The integration limits in the two different set of variables have to be transformed accordingly. The domain of integration of the first integral may thus depend on the physical variables in such a way that the integral is no longer a constant. Hence, not every G_v' is admissible[24]. We give an example in appendix 7.A. This problem does not arise in the version that we have used of the BRST derivation.

The Faddeev-Popov determinant can be expressed as an integral over $2kN$ Grassmann variables $\eta_v(t_a), \bar{\eta}_v(t_a)$, which become $2k$ fermion fields. They are called ghost fields since they are unphysical fermions which act as a device for calculations.

$$\begin{aligned}\Delta_{FP} &= \int \Pi_{a,v} d\eta_v(t_a) d\bar{\eta}_v(t_a)\, exp[\bar{\eta}_s(t_c)\{G_s(t_c), f_w(t_b)\}\eta_w(t_b)]]) \\ &\to \int \Pi_v D[\eta_v] D[\bar{\eta}_v]\, exp[\int \bar{\eta}_s(t)\Delta_{sw}(t,t')\eta_w(t')dtdt'] \qquad (5.11)\end{aligned}$$

If the G_v do not contain time derivatives, the Faddeev-Popov matrix is diagonal with respect to the time label so that

$$\begin{aligned}\Delta_{vw}(t_a, t_b) &\equiv \tilde{\Delta}_{vw}(t_a)\, \delta_{t_b t_a} \\ \to \Delta_{vw}(t,t') &\equiv \tilde{\Delta}_{vw}(t)\, \delta(t-t') \qquad (5.12)\end{aligned}$$

$$\Delta_{FP} = \int \Pi_v D[\eta_v][D\bar{\eta}_v]\, exp[\int \bar{\eta}_s(t)\tilde{\Delta}_{sw}(t)\eta_w(t)dt] \qquad (5.13)$$

5.2 The Coulomb or unitary gauge

Up to a constant, we may write

$$\Pi_v\, \delta(f_v) = lim_{D\to 0} exp[-i \int \frac{(f_v)^2}{2D} dt] \qquad (5.14)$$

In this case the gaussian must have an infinitely narrow width, since the constraints (unlike the gauge conditions) are not arbitrary. The combination of (5.6), (5.10), (5.13) and (5.14) yields the path integral

$$\begin{aligned}Z &= \lim_{D\to 0} \int \Pi_{v,i} D[I_v] D[\phi_v] D[p_i] D[q_i] D[\eta_v] D[\bar{\eta}_v]\, |\varsigma| \qquad (5.15) \\ & exp[i \int dt(p_j\dot{q}_j + P_w\dot{\phi}_w - i\bar{\eta}_s\tilde{\Delta}_{sw}\eta_w - \frac{1}{2A_w}G_w^2 - \frac{1}{2D}f_w^2 - H)]\end{aligned}$$

which is also obtained using an effective classical hamiltonian in the extended space of original, collective and ghost variables.

$$H_{eff} = \lim_{D\to 0} [H + \frac{1}{2D}f_v^2 + \frac{1}{2A_v}G_v^2 + i\bar{\eta}_v\tilde{\Delta}_{vw}\eta_w] \qquad (5.16)$$

---

[24] See also the comments below eq. (3.40).

Although the equivalence between H_{eff} and H has been demonstrated through a path-integral derivation, from now on we may proceed to solve the Schroedinger equation corresponding to H_{eff} within the usual quantal formalism. The hamiltonian operator is obtained by replacing in (5.16) the coordinates and momenta by the corresponding operators and the Poisson bracket $\{\ ,\ \}$ by $-i[\ ,\ ]$, where $[\ ,\ ]$ is the commutator.

The calculations using H_{eff} must be performed using a finite value of D. The limit $D \to 0$ should be taken only in the final physical results, which in this limit should be independent of the arbitrary constants A_v.

We may choose variables θ_v, which are (at least approximately) conjugate functions to J_v, as gauge fixing functions G_v. To the extent that this is possible the Faddeev-Popov determinant becomes a constant and the ghosts decouple from the remaining terms in H_{eff}. Moreover, the motion along the angular directions can be described by the harmonic hamiltonians

$$\hat{h}_v = \frac{1}{2D}\hat{J}_v^2 + \frac{1}{2A}\hat{\theta}_v^2 \tag{5.17}$$

which yield an infinite frequency in the limit $D \to 0$

$$\omega = \frac{1}{(AD)^{1/2}} \tag{5.18}$$

(we have taken all $A_v = A$).

Therefore, the collective motion is eliminated from the intrinsic degrees of freedom. However, the total number of degrees of freedom remains invariant, since the collective momenta I_v have appeared in the effective hamiltonian (cf. subsection 6.1.3). Because no spurious states are left, the exponentiation (5.15) of the constraint corresponds to the treatment in the physical or Coulomb gauge in field theory.

The hamiltonian displays a collective energy term $\hat{I}_v^2/2D$ and a Coriolis-type interaction $-\hat{I}_v\hat{J}_v/D$, both diverging in the limit $D \to 0$. The cancellation between these two contributions becomes apparent in a quantal perturbative calculation: in second order the Coriolis interaction yields an energy

$$-I_v^2| < 1|\hat{J}_v|0 > |^2/D^2w = -I_v^2/2D \tag{5.19}$$

where the value of the matrix element $< 1|\hat{J}_v|0 >= i(wD/2)^{\frac{1}{2}}$ has been obtained using the harmonic approximation (5.17).

There is no kinetic term for the Faddeev-Popov ghosts in the effective hamiltonian because the G_v do not include temporal derivatives. Therefore, the free ghost propagators are different from zero only at equal times and, consequently, a factor $\delta(0)$ appears in the corresponding Green's function. We make this explicit using an alternative procedure in order to exponentiate the Faddeev-Popov determinant, namely[25]

$$\begin{aligned}\Delta_{FP} &= exp[tr\ log\ \Delta_{vw}(t_a, t_b)] \\ &\rightarrow exp[\int \delta(t-t')\ log\ det[\Delta_{vw}(t,t')]\ dtdt'] \end{aligned} \tag{5.20}$$

In the first line of eq. (5.20) tr is taken both over the v, w indices and over the time indices t_a, t_b. In the second line (and in the following equations) the trace over the time indices has been explicitely taken and the det symbol applies only to the indices v, w.

In the particular case of a matrix diagonal in the time label,

$$\begin{aligned}\Delta_{FP} &= \int exp[\delta^2(t-t')\ log\ det[\tilde{\Delta}_{vw}(t)]\ dtdt'] \\ &= exp[\delta(0) \int\ log\ det[\tilde{\Delta}_{vw}(t)]\ dt] \end{aligned} \tag{5.21}$$

The det symbol applies here again to the $k.k$ matrix $\Delta_{vw}(t)$ We thus obtain a new effective hamiltonian

$$\begin{aligned}H_{eff} &= \lim_{D\rightarrow 0}[H + \frac{1}{2D}(I_v - J_v)^2 \\ &+ \frac{1}{2A}G_v^2 + i\ \delta(0)\ log\ \ det[\tilde{\Delta}_{vw}] \end{aligned} \tag{5.22}$$

to which all the discussion following eq.(5.16) applies as well. However we must assign a value to the constant $\delta(0)$. This may be accomplished through a regularization procedure using Ward identities and yields the value (see appendix 5.A). We obtain

$$i\delta(0) = -\omega/2 \tag{5.23}$$

where ω is given in eq.(5.18).

5.3 The Lorentz gauge

Another way of exponentiating a delta function is through a Fourier transform. Therefore the constraint (5.14) can also be expressed as

$$\Pi_v \delta(f_v) = \int \Pi_v D[\Omega_v] exp[i \int \Omega_w f_w dt] \tag{5.24}$$

This exponentiation of the constraints amounts to introduce the Lagrange multipliers Ω_v as new integration functions. In order to grant a kinetic energy to these degrees of freedom we proceed with the electromagnetic analogy

[25] We have used the continuous version of the expression $tr\ a_{ij} = a_{ij}\delta_{ij}$.

displayed in table 1 (or with an analogous Yang-Mills analogy for non-abelian transformations). In field theory, the Coulomb and Lorentz gauges are obtained with the gauge conditions

$$\begin{aligned} \partial_i A^i_v &= 0 \quad (i = 1,2,3) \\ \partial_i A^i_v - \dot{A}^o_v &= 0, \end{aligned} \tag{5.25}$$

respectively, where A^o_v are Lagrange multipliers. Therefore, from our previous choice of the gauge conditions $G_v = 0$ in the Coulomb case (G_v independent of Ω_v), we obtain "Lorentz" gauge conditions by imposing

$$G^{(L)}_v = \frac{G_v}{B_v} - \dot{\Omega}_v = 0 \tag{5.26}$$

where the arbitrary constants B_v have been included because there is no requirement of covariance in our cases. The factors $\delta(G^{(L)}_v)$ in the path-integral may be again exponentiated using the t'Hooft trick[26]

$$\Pi_v\ \delta(G^{(L)}_v) = exp[-i \int \frac{(B_v\dot{\Omega}_v - G_v)^2}{2A_v} dt] \tag{5.27}$$

The combination of (5.24) and (5.27) yields a lagrangian form of the path-integrals with respect to the Lagrange multipliers. The hamiltonian version is obtained through the replacement

$$\begin{aligned} exp[-i \int \frac{(B_v\dot{\Omega}_v - G_v)^2}{2A_v} dt] &= \text{constant}\ . \int \Pi_v D[P_v]\ exp[i \int (P_v\dot{\Omega}_v \\ &\quad - \frac{1}{B_v} P_v G_v + \frac{A_v}{2B^2_v} P^2_v) dt] \end{aligned} \tag{5.28}$$

where we have introduced the momenta P_v conjugate to the Lagrange multipliers Ω_v

$$P_v = -\frac{1}{2A_w} \frac{\partial (B_w\dot{\Omega}_w - G_w)^2}{\partial \dot{\Omega}_v} = \frac{B_v}{A_v}(G_v - B_v\dot{\Omega}_v) \tag{5.29}$$

(no summation over repeated indices in the r.h.s.). The inclusion of the new integration functions Ω_v, P_v in the path-integrals corresponds to perform the calculation in the **phase-space** $\mathcal{F}_3$ (cf. subsection 2.3). We notice that the replacement (2.84) has been implemented through the exponentiation of the constraints (eq. (5.24)).

Finally, we must exponentiate the Faddeev-Popov determinant. In $\mathcal{F}_3$ the gauge transformation takes the form (2.88). Consequently, the equation analogous to (5.8) is

$$\Delta_{vw}(t_a, t_b) = \frac{\partial}{\partial \beta_w(t_b)} T(\beta_s(t_c))\ G^{(L)}_v(t_a)|_{\beta_s = 0}$$

[26] We have replaced the constants A_v in (5.10) by A_v/B^2_v.

$$
\begin{aligned}
&= \frac{\partial}{\partial \beta_w(t_b)} \left(\dot{\beta}_s(t_c)\{P_s(t_c), \dot{\Omega}_v(t_a)\}\right. \\
&\quad \left. -c_{rsu}\beta_s(t_c)\Omega_u(t_c)\{P_r, \dot{\Omega}_v(t_a)\} + \frac{1}{B_v}\beta_s(t_c)\{f_s(t_c), G_v(t_a)\}\right) \\
&= -\frac{\delta_{vw}}{(t_{a+1}-t_a)^2}\left(\delta_{t_b t_{a+2}} + \delta_{t_b t_a} - 2\delta_{t_b t_{a+1}}\right) \\
&\quad +\frac{c_{vws}}{(t_{a+1}-t_a)}\Omega_s(t_b)\left(\delta_{t_b t_{a+1}} - \delta_{t_b t_a}\right) \\
&\quad -\delta_{t_a t_b}\frac{1}{B_v}\{G_v(t_a), f_w(t_b)\}
\end{aligned} \tag{5.30}
$$

As in the case of the Coulomb gauge, we have assumed that the functions G_v do not contain time derivatives. The exponentiation of the Faddeev-Popov determinant by means of ghosts now reads

$$
\begin{aligned}
\Delta_{FP} &= det[\Delta_{vw}(t_a, t_b)] \\
&= \int \Pi_{v.t_a} d\eta_v(t_a) d\bar{\eta}_v(t_a)\ exp[\bar{\eta}_r(t_c)\eta_w(t_b)\Delta_{rw}(t_c, t_b)] \\
&= \int \Pi_{v.t_a} d\eta_v(t_a) d\bar{\eta}_v(t_a)\ exp[(\bar{\eta}_w(t_b)\frac{d^2\eta_w(t_b)}{dt_b^2} \\
&\quad -c_{rws}\Omega_s(t_b)\dot{\bar{\eta}}_r(t_b)\eta_w(t_b) - \frac{1}{B_v}\bar{\eta}_r(t_b)\eta_w(t_b)\{G_v(t_b), f_w(t_b)\})]
\end{aligned} \tag{5.31}
$$

which yields in the continuous-time limit

$$
\begin{aligned}
\Delta_{FP} &= \int D[\eta_v]D[\bar{\eta}_v] \\
&\quad exp[\int dt\ (\dot{\eta}_w\dot{\bar{\eta}}_w - c_{rws}\Omega_s\dot{\bar{\eta}}_r\eta_w - \frac{1}{B_r}\bar{\eta}_r\eta_w\{G_r, f_w\})]
\end{aligned} \tag{5.32}
$$

We are again interested in the phase-space version of this path-integral, namely

$$
\Delta_{FP} = \int \Pi_v D[\pi_v]D[\eta_v]D[\bar{\pi}_v]D[\bar{\eta}_v]\ exp[i\int L_{gh}dt] \tag{5.33}
$$

Again, by performing the opposite operation of "integrating over momenta" (but for the ghost variables), we obtain a hamiltonian version of the ghost lagrangian

$$
\begin{aligned}
L_{gh} &= i(\dot{\bar{\eta}}_v\dot{\eta}_v + c_{vws}\Omega_s\dot{\bar{\eta}}_v\eta_w + \frac{1}{B_v}\bar{\eta}_v\eta_w\{G_v, f_w\}) \\
&= -\pi_v\dot{\eta}_v - \bar{\pi}_v\dot{\bar{\eta}}_v - H_{gh} \\
H_{gh} &= i\pi_v\bar{\pi}_v + c_{vws}\Omega_s\pi_v\eta_w - i\frac{1}{B_v}\bar{\eta}_v\eta_w\{G_v, f_w\} \\
\pi_v &= \frac{\partial L_{gh}}{\partial\dot{\eta}_v} = -i\dot{\bar{\eta}}_v
\end{aligned}
$$

$$\bar{\pi}_v = \frac{\partial L_{gh}}{\partial \dot{\bar{\eta}}_v} = i\dot{\eta}_v + ic_{vws}\Omega_s\eta_w \qquad (5.34)$$

The combination of eqs. (5.24), (5.28) and (5.34) yields the effective classical hamiltonian

$$\begin{aligned} H_{eff} &= H - \Omega_v f_v + \frac{1}{B_v} P_v G_v - \frac{A_v}{2B_v^2} P_v^2 \\ &\quad + i\pi_v \bar{\pi}_v + c_{vws}\Omega_s \pi_v \eta_w - i\frac{1}{B_v}\bar{\eta}_v \eta_w \{G_v, f_w\} \,. \end{aligned} \qquad (5.35)$$

The BRST hamiltonian (4.18) is obtained by substituting the classical coordinates and momenta by their corresponding operators, $\{\ ,\ \} \to -i[\ ,\ ]$ and $c_{vws} \to ic_{vws}$.

There are apparent formal differences between the Faddeev-Popov procedure and the method based on the BRST invariance. In the former case we deal with a singular lagrangian with a local gauge invariance. In the second case we use a regular lagrangian with a global (supersymmetric) invariance. However the result (5.35) verifies their equivalence.

appendix 5.A The value of $i\delta(0)$

The value of $i\delta(0)$in the hamiltonian H_{eff} (eq. (5.22)) may be derived through the use of Ward identities. We add a source term corresponding to the functions G_v in the path-integral

$$Z(M_s) = \int D[\tau]\, exp[i \int (L_{eff} + M_v G_v] \tag{5.36}$$

where

$$\begin{aligned} L_{eff} &= p_i \dot{q}_i + I_v \omega_v - H_{eff} \\ D[\tau] &= \Pi_{i,v}\ D[p_i] D[q_i] D[I_v] D[\phi_v]\ |\varsigma| \end{aligned} \tag{5.37}$$

The amplitude $Z(M_s)$ is invariant under the gauge (canonical) transformation (2.83). The only terms that are affected are those depending on the G_v,

$$\begin{aligned} 0 &= i \int D[\tau]\ \left(-\frac{1}{2A}\{f_v(t_a), G_w^2(t_a)\} + M_w(t_a)\{f_v(t_a), G_w(t_a)\}\right) \\ & exp[i \int (L_{eff} + M_s G_s)dt] \end{aligned} \tag{5.38}$$

As in eq. (5.17) we assume that the G_v can be approximated by the variables θ_v that are conjugate to the J_v ($\{f_v, \theta_w\} = -\delta_{vw}$). Therefore,

$$0 = \left(-\frac{1}{A} \frac{\partial}{\partial M_v(t_a)} + iM_v(t_a) \right)\ Z(M_s) \tag{5.39}$$

This is a Ward identity. The functional differential equation may be immediately solved:

$$Z(M_s) = Z(0)\ exp[i \frac{A}{2} \int M_v^2(t)dt] \tag{5.40}$$

The first derivative yields the vanishing result

$$0 = \frac{\partial Z(M_s)}{\partial M_v(t_a)}|_{M_s=0} = i \int D[\tau]\ \theta_v(t_a)\ exp[i \int L_{eff} dt] \tag{5.41}$$

which says that the expectation value of θ_v vanishes. This is consistent with our previous treatment of the functions G_v (eq. (5.10)). Moreover,

$$\begin{aligned} \frac{\partial^2 Z(M_s)}{\partial M_v(t_a) \partial M_w(t_b)}|_{M_s=0} &= iZ(0)\ A\ \delta_{vw} \delta(t_a - t_b) \\ &= -\int D[\tau]\ \theta_v(t_a)\theta_w(t_b)\ exp[i \int L_{eff} dt] \end{aligned} \tag{5.42}$$

This equation may be used to obtain $i\delta(0)$ within the harmonic approximation (5.17)

$$\begin{aligned} i\delta(0) &= -\frac{A}{Z(0)}\int D[\tau]\,\theta_v^2(t_a)\,exp[i\int L_{eff}dt] \\ &= -A < \theta_v^2 >= -\frac{1}{2}\omega \end{aligned} \qquad (5.43)$$

6 The breaking and the restoration of symmetries

There are many instances in which the original symmetries which are present in the hamiltonian disappear from the solution describing (approximately) the ground state, either classically or quantally. More precisely, the approximate solutions do not belong to an irreducible representation of the symmetry group. In these cases the "vacuum" solution may be transformed into another solution by application of the transformation operator associated with the original symmetries. Obviously the new solutions have the same energy as the original vacuum. In other words, the approximate vacua are degenerate, which manifests itself by the appearance of modes with zero frequency (at least for continuous groups of transformations). If the symmetry breaking is only approximate, these zero modes yield a bad description of the motion associated with the symmetry group. Moreover, naive perturbation theory is no longer possible due to the existence of infrared singularities[27].

Around 1960 Nambu and Goldstone realized the significance of this notion of symmetry breaking in condensed matter physics. In 1964 Higgs pointed out that the consequences of symmetry breaking in gauge theories are different from those in non-gauge theories.

6.1 A simple model: the Mexican hat potential

We first discuss some relevant aspects of the (approximate) breaking of symmetries within the framework of a very simple model. The generalizations are outlined at the end of the section.

Let us consider the planar motion of a particle subject to the "Mexican hat" potential

$$\begin{aligned} H &= \frac{1}{2m}(p_1^2+p_2^2)+V(r) \; ; \; r^2=|z|^2=q_1^2+q_2^2 \\ V(r) &= \frac{m\omega_x^2}{8|r_o^2|}(r^2-r_o^2)^2 \end{aligned} \tag{6.1}$$

which depends on two parameters, ω_x and r_o. The potential is depicted in fig. 1. This classical hamiltonian has a rotational (cylindrical) invariance associated with the angular momentum generator J (2.35).

[27]Even if there is no symmetry breaking, the vacuum may still belong to an irreducible representation different from the invariant representation. For instance, the vacuum of an atom with an odd number of electrons carries a semiinteger angular momentum. In such cases the group transformation takes the system from one substate to another substate associated with the same irreducible representation. However, from the point of view of perturbation theory, the vacuum behaves as being non-degenerate, since there is no way to connect these two substates with a scalar operator like the hamiltonian.

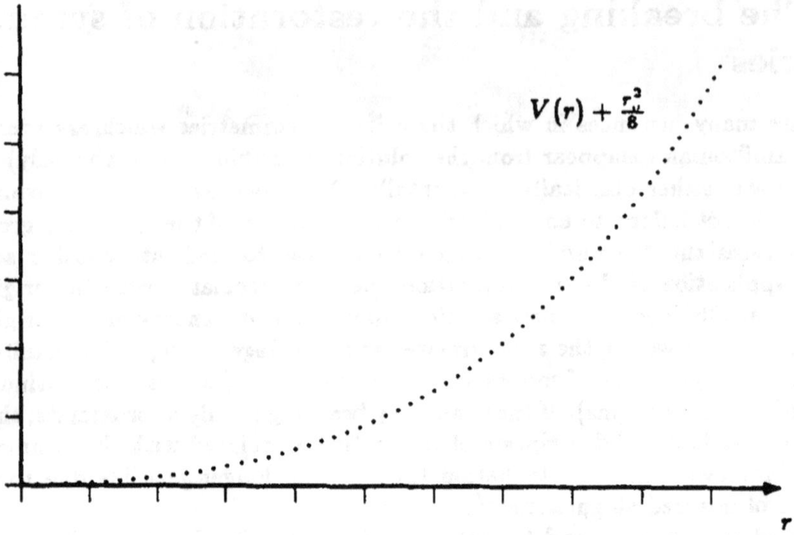

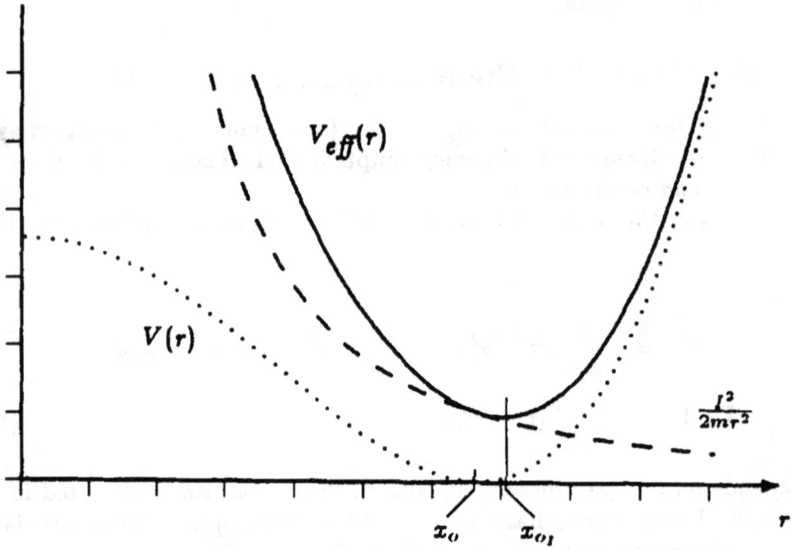

Figure 1: The Mexican hat potential. The dotted line represents the potential (6.1), the dashed line the centrifugal potential and the full line the effective potential. The relation between r_o and x_o is given in eq. (6.5)

If $r_o^2 < 0$, the classical vacuum of the system is given by $(q_{1o}, q_{2o}) = (0, 0)$. This vacuum is left invariant by the symmetry transformation. There are two normal modes of oscillation around the minimum. They have both the same frequency $\omega_x/\sqrt{2}$.

If $r_o^2 > 0$, there is a continuous family of minima such that

$$q_{1o}^2 + q_{2o}^2 = r_o^2 \tag{6.2}$$

Because of the existence of the continuous family of minima (6.2) we may choose, for instance, a vacuum such that the particle lies on the 1-axis ($z_o = real = x_o$). Therefore, although the system possesses a cylindrical symmetry, the ground state ($q_{1o} = x_o$; $q_{2o} = 0$) is not invariant under that symmetry. Rather, it changes under rotations into one of the other equivalent (and degenerate) ground states in which the vector position of the particle makes a finite angle with the 1-axis. This example exhibits a spontaneous breaking of symmetry: the symmetry of the hamiltonian is not shared by the approximate (classical) ground state solution.

If we consider small oscillations around these minima we find that there is a restoring force along the radial direction, since it costs energy to displace the particle against the potential. On the contrary, there is no restoring force in the direction of the valley. Thus, the normal mode corresponding to the angular degree of freedom has zero frequency. From the quantum mechanical point of view, this brings in the problem that we cannot perturb around a classical vacuum (q_{1o}, q_{2o}) using a harmonic oscillator basis: because of the presence of a zero frequency mode, the oscillations associated with this mode become degenerate with the ground state.

The appearance of a zero mode associated with rotations suggests that we should try to single out the "angular" variable. In this simple example this is very easy to do by resorting to polar coordinates which handy separate the rotational motion from the radial degree of freedom.

6.1.1 The separation of variables

As mentioned in the introduction, a technique of separation of the most relevant variables (in this case the rotational variable) is the most straightforward method to use. Unfortunately, it is generally not feasible in practice, even in cases when it is theoretically possible. In this simple example we use it to compare with less straightforward methods which, however, may be also applied to realistic models.

Let us first rewrite the quantum version of the hamiltonian (6.1) ($r_o^2 > 0$)

$$\hat{H} = \frac{1}{2m}(\hat{p}_1^2 + \hat{p}_2^2) + \frac{m\omega_x^2}{8r_o^2}(\hat{r}^2 - r_o^2)^2 \tag{6.3}$$

where

$$[\hat{q}_i, \hat{p}_j] = i\hbar\delta_{ij} \tag{6.4}$$

For convenience we adimensionalise the variables and parameters as follows

$$\begin{aligned}
\hat{p}_i &= \frac{1}{\sqrt{m\hbar\omega_x}}\hat{\hat{p}}_i \\
\hat{q}_i &= \sqrt{\frac{\omega_x m}{\hbar}}\hat{\hat{q}}_i \\
x_o &= \sqrt{\frac{\omega_x m}{\hbar}} r_o
\end{aligned} \tag{6.5}$$

so that, as in previous sections

$$[\hat{q}_i, \hat{p}_j] = i\delta_{ij} \tag{6.6}$$

and $\hat{\hat{H}}$ becomes

$$\hat{H} = \frac{1}{\hbar\omega_x}\hat{\hat{H}} = \frac{\hat{p}_i^2}{2} + \frac{1}{8x_o^2}(\hat{r}^2 - x_o^2)^2 \tag{6.7}$$

We write $\hat{H}$ in polar coordinates [30]

$$\hat{H} = \frac{\hat{p}_r^2}{2} + \frac{(\hat{J}^2 - 1/4)}{2\hat{r}^2} + \frac{1}{8x_o^2}(\hat{r}^2 - x_o^2)^2 \tag{6.8}$$

where this hamiltonian acts on the wave functions redefined as $\Phi \rightarrow r^{-1/2}\Phi$.

Working within a given I-subspace we obtain the radial hamiltonian

$$\begin{aligned}
\hat{H} &= \hat{H}_I |I><I| \\
\hat{H}_I &= \frac{\hat{p}_r^2}{2} + \frac{(I^2 - 1/4)}{2\hat{r}^2} + \frac{1}{8x_o^2}(\hat{r}^2 - x_o^2)^2
\end{aligned} \tag{6.9}$$

Here we have trivially separated (*not* uncoupled) the rotations, and we are left with a single vibrational degree of freedom. The potential has a "wall" at $r \rightarrow 0$, which only has an exponentially small effect for sufficiently large values of x_o.

From the last of eqs. (6.5) we recognize x_o as a semiclassical parameter. This parameter plays the role of the symmetry breaking parameter in field theory, of the deformation parameter in nuclear physics and of the order parameter in solid state physics.

The presence of the centrifugal term $I^2/2\hat{r}^2$ yields an effective potential which is also represented in fig. 1. The minimum solution is given by the equation

$$\frac{d}{dr}\left(\frac{I^2}{2r^2} + \frac{1}{8x_o^2}(r^2 - x_o^2)^2\right)\Big|_{r=x_{oI}} = 0$$

$$x_{oI} = x_o\left(1 + \frac{2I^2}{x_o^2}\right) \tag{6.10}$$

We may expand $\hat{H}_I$ in inverse powers of x_{oI} as follows

$$
\begin{aligned}
\hat{H}_I &= \hat{H}_{vib} + \hat{H}_{rot} + \hat{H}_{anh} + \hat{H}_{coup} \\
\hat{H}_{vib} &= \frac{1}{2}(\hat{p}_r^2 + \hat{r}^2) \\
\hat{H}_{rot} &= \frac{I^2}{2\Im_I}; \quad \Im_I = x_{oI}^2 \\
\hat{H}_{anh} &= (\frac{\hat{r}^3}{2x_{oI}} + \frac{\hat{r}^4}{8x_{oI}^2} + ...) \\
\hat{H}_{coup} &= \frac{3I^2\hat{r}^2}{\Im_I x_{oI}^2} + ...
\end{aligned}
\tag{6.11}
$$

For large values of x_{oI} the system becomes a vibrator plus a rotor with an I-dependent moment of inertia. The rotor even becomes completely rigid in the limit $I^2 << x_o$. There are also smaller anharmonicities and vibration-rotation couplings. The spectrum is represented in fig. 2.

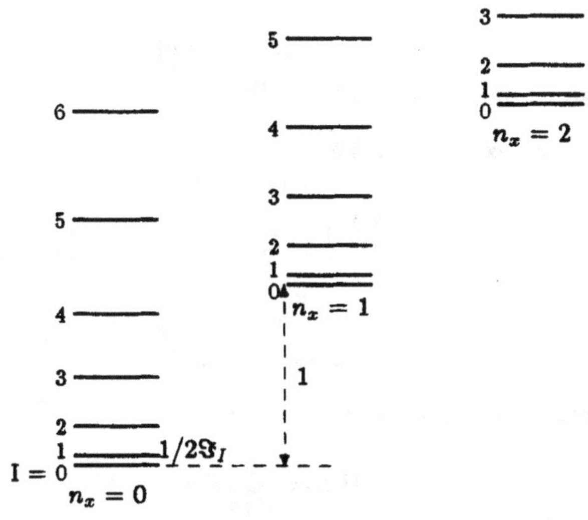

Figure 2: The spectrum of the hamiltonian with a Mexican hat potential for large values of x_{oI}

The energy of a vibrational quantum divided by a typical separation energy between rotational states belonging to the same band is

$$\frac{E_{vib}}{E_{rot}} = \frac{1}{1/2\Im_I} = x_{oI}^2 \tag{6.12}$$

Hence the approximation of a stable minimum also implies that the vibrational energy (associated with fluctuations in the deformation) is large compared with the energy associated to the rotations. This "degeneracy" between the states belonging to the same rotational band (relative to vibrational excitations) is consistent with the fact that the rotation is a free motion without restoring force, the quantization arising only through requirements of periodicity[28].

6.1.2 The symmetry breaking approximations

As we mentioned in the Introduction, the separation of variables performed in the preceding paragraphs is not easily implemented in realistic many-body problems. Hence, the usual practice is to use symmetry-breaking approximations and (subsequently) to restore the symmetry (at least approximately).

Consider the family of (non-normalisable) approximate vacua

$$|q_1\, q_1 > \propto \delta(q_1 - q_{1o})\, \delta(q_2 - q_{2o}) \quad ; \quad q_{1o}^2 + q_{2o}^2 = x_o^2 \tag{6.13}$$

and in particular

$$|x_o\, 0 > \propto \delta(q_1 - x_o)\, \delta(q_2) \tag{6.14}$$

This state can be written in polar coordinates as

$$\begin{aligned} |x_o\, 0 > &\propto \delta(r - x_o)\, \delta(\phi) \\ &\propto \delta(r - x_o) \left[\sum_{I=-\infty}^{I=+\infty} e^{iI\phi}\right] \propto |x_0 > \left[\sum_{I=-\infty}^{I=+\infty} |I>\right] \end{aligned} \tag{6.15}$$

where we have defined the "radial" state

$$|x_o > \propto \delta(r - x_o) \tag{6.16}$$

and the angular momentum eigenfunctions

$$|I> = \frac{1}{\sqrt{2\pi}} e^{iI\phi} \tag{6.17}$$

We conclude that the approximate vacuum (6.15) is a combination of the whole rotational band associated with the approximate radial vacuum $|x_o>$. This is a

[28] Had we considered translational instead of rotational symmetries, we would have obtained a continuous band of translational states. There are no requirements of periodicity in such a case.

quite general fact: a quantum solution which is characterized by a well defined angle (i.e., a wave packet confined to a particular classical vacuum (q_{1o}, q_{2o})) looses the quantum number corresponding to the angular momentum. Thus, a state with broken symmetry can be considered as a combination of the whole rotational band associated with a given vibrational state.

Clearly, these symmetry-breaking solutions centered around a particular deformed vacuum can never be exact, unless the band associated with the motion "along the symmetry group" becomes degenerate. This can only happen quantum mechanically in infinite systems; for example when the inertial parameter is proportional to the volume. In finite systems, however, it may take place as a semiclassical limit. Such is the case if we assume x_o to be large.

6.1.3 The collective coordinates

Consider the system associated with the hamiltonian (6.1) expressed in a rotating frame described by an angle ϕ. We write the rotating lagrangian (2.20) in terms of the fluctuations

$$L = \frac{m}{2}\dot{q}_1^2 + \frac{m}{2}\dot{q}_2^2 + mr_o\omega\dot{q}_2 + \frac{m}{2}r_o^2\omega^2 - \frac{m\omega_x^2}{2}(q_1 - r_o)^2 + \text{cubic terms} + ... \quad (6.18)$$

The third term in the r.h.s. suggests that the ϕ and the q_2-degrees of freedom may turn into each other. In fact, making use of the gauge transformation (2.29), we may choose $\delta\alpha$ so as to make $\dot{q}_2 = 0$. In this (unitary) gauge, the lagrangian becomes

$$L = \frac{m}{2}\dot{q}_1^2 + \frac{m}{2}r_o^2\omega^2 - \frac{m\omega_x^2}{2}(q_1 - r_o)^2 + \text{cubic terms} + ... \quad (6.19)$$

The q_2-boson, which in the case of spontaneous breaking of the global symmetry has zero frequency (Goldstone boson), has disappeared. Simultaneously, a finite frequency is associated with the gauge degree of freedom, whose existence is due to the presence of a local gauge symmetry (cf. subsection 2.1). The corresponding motion as determined by the term $\frac{1}{2}\Im\omega^2$ represents a rotation with a moment of inertia as in (6.11).

This is analogous to the Higgs phenomenon: in gauge field systems for abelian models, the breaking of a symmetry does not result in the presence of a massless Goldstone boson, but in the disappearance of that field and the appearance, instead, of a massive gauge field.

However although both in (6.19) and in field theory the gauge field has acquired a mass, the meaning of this expression is different in the two cases. In field theory there appears a constant (the mass) multiplying the square of the gauge field. This would be read as a restoring force in the appropriate direction for a mechanical problem. On the contrary, the term $\frac{1}{2}\Im\omega^2$ (which is also a constant time the square of the gauge field) is a kinetic energy, due to the fact that the mechanical gauge field is the time-derivative of the collective coordinate (cf. subsection 2.1).

The same ideas can be applied quantum mechanically[29] using the Coulomb gauge of subsection 5.2. We choose as above a gauge-fixing function $G \propto q_2$. The factor $\delta(G)$ which appears in the path integral (or the corresponding term $\frac{G^2}{2A}$ in the effective hamiltonian) fixes the gauge. Although for finite A a spurious non-zero energy mode survives, perturbation theory becomes feasible. The spurious modes disappear from the spectrum if the constraint is enforced by taking the limit $D \to 0$.

Within the BRST treatment the projection as well as the gauge-fixing mechanisms are more subtle. It is worthwhile reviewing them in some detail. We choose the gauge fixing function $\hat{\rho}$ as in subsection 4.1

$$\hat{\rho} = \hat{\Omega}\hat{\pi} + \hat{\eta}(\frac{1}{B}\hat{G} - \frac{A}{2B^2}\hat{P}) \tag{6.20}$$

(with $\hat{G}$ as above) so that the effective hamiltonian becomes

$$\begin{aligned} \hat{H}_{BRST} &= \hat{H} + [\hat{\rho}, \hat{Q}]_+ \\ &= \hat{H} + i\hat{\pi}\hat{\pi} - \hat{\Omega}\hat{f} + \frac{1}{B}\hat{G}\hat{P} - \frac{A}{2B^2}\hat{P}^2 \\ &\quad - \frac{1}{B}\hat{\bar{\eta}}[\hat{G}, \hat{f}]\hat{\eta} \end{aligned} \tag{6.21}$$

The advantages of $\hat{H}_{BRST}$ with respect to $\hat{H}$ are the following:

---

[29]There are several alternative treatments that are based on the deformed "classical" solution and on some procedure that restores the broken symmetry (and hence generates the band associated with that symmetry). Some of these methods use the operator that projects onto an irreducible representation of the group. For instance, consider the projection operator

$$P_I = |I><I|$$

By applying it to the deformed vacuum (6.14) and by using expression (6.15), we obtain

$$|\Psi^I> \equiv P_I|x_o, 0> = |x_o>|I>$$

The expectation values of the hamiltonian with this wavefunction can be calculated by noticing that

$$\begin{aligned} P_I\hat{H}P_I &= \hat{H}_I|I><I| \\ <\hat{H}>_{x_o} &= V(x_o) + \frac{I^2}{2x_o^2} \end{aligned}$$

(we are neglecting the term $-\frac{1}{4x^2}$ of eq.(6.8)). We have performed the classical minimization *before* projecting, and hence we have not taken into account centrifugal effects. A more exact procedure is to project first and minimize afterwards. Clearly, this is equivalent to minimizing H_I with radial wave functions. The minimization equation is therefore eq. (6.10) and takes into account centrifugal effects. In both cases one obtains a rotational band associated with the approximate (classical) solution.

These projection procedures have been applied extensively in nuclear physics. We only discuss them here for comparison purposes, since our aim is to treat the broken symmetries with collective coordinates.

i) $\hat{H}_{BRST}$ commutes with the collective operator $\hat{I}^2$. Therefore, $\hat{H}_{BRST}$ displays an unbroken (collective) symmetry. The corresponding wave functions are labelled by the quantum number I associated with the group of transformations generated by the collective operators.

$$|\Psi> = |I> \otimes |intr> \tag{6.22}$$

where the rotational wave functions $|I>$ are given in eq.(6.17) and $|intr>$ depends on q_1, q_2, Ω and on the ghosts.

ii) $\hat{H}_{BRST}$ no longer commutes with the generators $\hat{J}$ (as $\hat{H}$ does) due to the presence of the gauge-fixing functions $\hat{G}$. Therefore infrared divergences are eliminated and we can perform perturbative calculations.

iii) The projection over an irreducible representation of the original group of transformations is automatically achieved. First we note that

$$\hat{J}^2 = \hat{I}^2 - [\hat{Q}, \hat{\pi}(\hat{J} + 2\hat{I})]_+ \tag{6.23}$$

so that $\hat{J}^2$ differs from $\hat{I}^2$ only to within an $\hat{O}_\chi$ (null) operator. We may construct a state $|I, n>$, labelled by the representation of the group plus additional quantum numbers n, which is an eigenstate of H_{BRST}, of $\hat{I}^2$ and is annihilated by $\hat{Q}$ (these three operators commute). Therefore,

$$\begin{aligned} \hat{J}^2|I, n> &= (\hat{I}^2 + \hat{O}_\chi)|I, n> \\ &= I^2|I, n> + |\chi> \end{aligned} \tag{6.24}$$

so that $|I, n>$ differs from an eigenstate of $\hat{J}^2$ by a zero-norm $|\chi>$-state, which is irrelevant as far as overlaps are concerned. Therefore, the state $|I, n>$ is for all practical purposes a projected state. In realistic cases we may work with states that are strictly eigenstates of $\hat{I}^2$ (due to the form of the hamiltonian) but are only approximately annihilated by $\hat{Q}$. To the extent that the states become more physical by improving the condition $\hat{Q}|I, n> = 0$, they also become more projected in the sense of (6.24).

The fact that the wave functions are not strictly eigenfunctions of $\hat{Q}$ does not produce any problem from the point of view of perturbation theory, since this approximately broken supersymmetry does not bring in zero modes.

The results of the semiclassical minimization of H_{BRST} within a I-subspace are given in eq. (7.1). We note that the vacuum is unique, i.e. we do not have a continuous family of minima (thanks to gauge-fixing). Finite size effects are incorporated perturbatively through an expansion in powers of x_{oI}^{-1}.

6.2 Some more general considerations

The physical considerations which we outlined in the previous subsections generalize in a number of ways.

Firstly, we consider usually broken symmetries associated with larger groups, SU(2) being a typical case (see section 8). In these cases the vacuum may be still invariant with respect to a subgroup T_h of the group of symmetries T_g ($T_h \epsilon T_g$) of the hamiltonian. There are degrees of freedom for which the restoring force is not required to vanish (although it may "by accident"). Their number is given by the dimension of the algebra associated with T_h. The elements of T_g which do not belong to T_h form a set T_g/T_h (they do not form a group, since the identity belongs to T_h). The number of Goldstone bosons is the dimension of the coset space, which is the number of generators of T_g that are not also generators of T_h.

In many-fermion systems the Hartree (Hartree-Fock, Hartree-Fock-Bogoliubov) vacuum can also be considered as classical vacua (although not the standard $\hbar \rightarrow 0$ case). This can be seen for example in that they can be derived as stationary phase approximations of the coherent-state path integrals, and in the related context of bosonizations [31]. The small oscillations around such vacua are obtained through the random-phase approximation (RPA). Consequently, most of the discussion in this section carries over straightforwardly to many-fermion or many-boson systems.

7 The application of the BRST formalism to abelian transformations

In the first place we discuss in some detail the application of the BRST formalism to the problem discussed in section 6. The advantages of having a simple model displaying the essential physics of more complicated situations have been systematically exploited in physics. In our case we thus achieve two purposes, namely: i) to understand how the system works through a simple example and ii) to develop a procedure for solving the BRST hamiltonian which may be readily extended to more complicated situations. In subsection 7.2 we show how this generalization can be carried out for the case of a many-body problem which admits a BCS description. In appendix 7.A we solve again the simple problem using the formalism derived for the Coulomb gauge in section 5 and we discuss limitations on the gauge conditions appearing in such case.

7.1 The Mexican hat potential

As for many physical problems, we divide the solution into three steps: i) the variational procedure yielding the ground state solution; ii) the determination of the normal modes; and iii) the calculation of the perturbative corrections. In systems displaying a breakdown of symmetries, this last step cannot be carried out using normal perturbation theory with the original hamiltonian, due to the presence of zero frequency modes (section 6). The BRST formalism eliminates these difficulties.

7.1.1 The variational procedure

According to eq. (6.22) we may work within a given I-subspace. Assuming $G \propto q_2$ and neglecting the coupling with the ghosts, a classical minimization of H_{BRST} (eq. (6.21)) with respect to the variables (q_i, p_i, Ω, P) yields the equilibrium values:

$$\begin{aligned} <\hat{P}> &= <\hat{q}_2> = <\hat{p}_1> = 0 \\ <\hat{p}_2> &= \frac{I}{x_{oI}}; \quad <\hat{\Omega}> = \frac{I}{x_{oI}^2} \end{aligned} \tag{7.1}$$

and the expectation value x_{oI} for $<\hat{q}_1>$ (eq. (6.10)).

In addition to the frequency of the radial oscillation (which has been eliminated through the transformation (6.5) to adimensional variables), there are still two physical parameters in this model. The first one can be taken to be x_{oI}. Clearly x_{oI} is also a semiclassical parameter because the fluctuations around the mean value (6.10) vanish in the classical limit. Therefore, an expansion in powers of x_{oI}^{-1} becomes appropriate if x_{oI} is sufficiently large. The second independent parameter can be taken to be either the average angular frequency

$<\hat{\Omega}>$ or the angular momentum I. However, the calculation simplifies if we assume that I (and not $<\hat{\Omega}>$) is of order unity. In the following we assume that this is the case. Consequently, we use $x_o^{-1} = x_{o,I=0}^{-1}$ as our expansion parameter

We express the hamiltonian and other operators in terms of the equilibrium values (7.1) and of the fluctuations around them. The fluctuations are hereafter labelled by $\hat{q}_i, \hat{p}_i, \hat{\Omega}$ and $\hat{P}$ (as previously the total operators). For example, the generator $\hat{J}$ has the form

$$\begin{aligned}\hat{J} &= (x_{oI} + \hat{q}_1)(\frac{I}{x_{oI}} + \hat{p}_2) - \hat{x}_2\hat{p}_1 \\ &= I + x_{oI}\hat{p}_2 + \frac{I}{x_{oI}}\hat{q}_1 + \hat{q}_1\hat{p}_2 - \hat{q}_2\hat{p}_1 \qquad (7.2)\end{aligned}$$

Therefore, the largest term is linear in the boson variables, namely

$$\hat{J}^{(1)} = x_o\hat{p}_2 = \Im^{1/2}\hat{p}_2 \qquad (7.3)$$

where $\Im$ is the moment of inertia given in eqs. (6.11) in the limit $I >> x_o^2$. It is convenient to define the gauge fixing operator $\hat{G}$ as the conjugate variable (at least to leading order in x_o^{-1})

$$\hat{G}^{(1)} = \frac{1}{\Im^{1/2}}\hat{q}_2 \qquad (7.4)$$

$$[\hat{G}^{(1)}, \hat{J}^{(1)}] = i \qquad (7.5)$$

The vanishing of the expectation value of $\hat{G}^{(1)}$ (cf. eq. (7.1)) is consistent with eq. (4.21). This condition fixes the moving frame of reference with respect to the system (particle) in such a way that the particle moves along the q_1-axis.

7.1.2 The normal modes

The hamiltonian $\hat{H}^{(2)}$, up to quadratic order in the fluctuations, may be divided into a term describing real degrees of freedom, a spurious boson term and a ghost term. We consistently retain in $\hat{H}^{(2)}$ only the leading order terms in the expansion in powers of x_o^{-1}

$$\hat{H}^{(2)} = \hat{H}_{re}^{(2)} + \hat{H}_{sp}^{(2)} + \hat{H}_{gh}^{(2)} \qquad (7.6)$$

i) the real degrees of freedom

$$\begin{aligned}\hat{H}_{re}^{(2)} &= \frac{I^2}{2\Im} + \frac{1}{2}(\hat{p}_1^2 + \hat{q}_1^2) \\ &= \frac{I^2}{2\Im} + \Gamma_x^\dagger\Gamma_x + \frac{1}{2} \qquad (7.7)\end{aligned}$$

where the creation operator of the real phonon is defined as

$$\Gamma_x^\dagger = \frac{1}{\sqrt{2}}(\hat{q}_1 - i\hat{p}_1) \tag{7.8}$$

In eq. (7.7), the first degree of freedom (ϕ, I) gives rise to a collective rotational energy and is an analogue to the Higgs phenomenon. As mentioned before, in a gauge system the breaking of a symmetry does not result in the appearance of a massless Goldstone boson but of a massive gauge field. The second degree of freedom $(\Gamma_x^\dagger, \Gamma_x)$ represents real vibrations along the q_1-axis.

ii) The term with spurious bosons

$$\hat{H}_{sp}^{(2)} = \frac{1}{2\Im}(\hat{J}^{(1)})^2 - \frac{A}{2B^2}\hat{P}^2 - \hat{\Omega}\hat{J}^{(1)} + \frac{1}{B}\hat{P}\hat{G}^{(1)} \tag{7.9}$$

This hamiltonian represents two coupled oscillators with zero restoring force. The corresponding normal modes are created by the operators $\Gamma_n^\dagger$ $(n = 0, 1)$ through the transformation[30]

$$\begin{aligned}
\hat{J}^{(1)} &= \Im^{1/2}\hat{p}_2 = (\Im\omega/2)^{1/2}(\Gamma_1^\dagger + \Gamma_1 + \Gamma_o^\dagger + \Gamma_o)\\
\hat{G}^{(1)} &= \hat{q}_2/\Im^{1/2} = -i(2\Im\omega)^{-1/2}(\Gamma_1^\dagger - \Gamma_1)\\
\hat{P} &= -i(\Im/2\omega)^{1/2}(\Gamma_1^\dagger - \Gamma_1 + \Gamma_o^\dagger - \Gamma_o)\\
\hat{\Omega} &= (\omega/2\Im)^{1/2}(\Gamma_o^\dagger + \Gamma_o)
\end{aligned} \tag{7.10}$$

It is important to note that the operators Γ_o, Γ_1 *are not* the same as those defined in subsection 4.2.2. In fact, neither the collective coordinate $\hat{\phi}$ nor the associate momentum $\hat{I}$ appear in the quadratic hamiltonian (7.9). The degrees of freedom entering in $\hat{H}_{sp}^{(2)}$ are the spurious sector of the particle motion and the Lagrange multiplier. For this last one we may use the metric discussed in subsection 3.2 and thus a transformation formally identical to that used in subsection 4.2.2 can be subsequently applied. Therefore, this definition of the metric of the phonons Γ_o, Γ_1 is used in this section and in the following ones.

The two bosons $\Gamma_n^\dagger$ have the same frequency ω

---

[30] The two degrees of freedom are uncoupled by a transformation similar to the one in eq. (7.10), $\hat{G}^{(1)}$ playing the same role as $\hat{\phi}$ did before. This is equivalent to complete the squares in (7.9)

$$\hat{H}_{sp}^{(2)} = \frac{1}{2\Im}(\hat{J}^{(1)} - \Im\hat{\Omega})^2 - \frac{A}{2B^2}(\hat{P} - \frac{B}{A}\hat{G}^{(1)})^2 + \frac{1}{2A}(\hat{G}^{(1)})^2 - \frac{\Im}{2}\hat{\Omega}^2$$

where the ratio (7.12) can be determined through the uncoupling condition

$$[(\hat{J}^{(1)} - \Im\hat{\Omega})\,,\,(\hat{P} - \frac{B}{A}\hat{G}^{(1)})] = 0$$

$$\omega = 1/\sqrt{B} \tag{7.11}$$

which is one of our arbitrary parameters. Special requirements appear for the diagonalizability of quadratic expressions having degenerate normal modes. As a consequence the value of A becomes fixed, namely

$$A = B/\Im \tag{7.12}$$

More importantly, the mode with $n = 0$ has a negative metric (cf. eq. (4.44))

$$[\Gamma_1, \Gamma_1^\dagger] = [\Gamma_o^\dagger, \Gamma_o] = 1 \tag{7.13}$$

The uncoupled expression for $\hat{H}_{sp}^{(2)}$ is, therefore

$$\hat{H}_{sp}^{(2)} = \omega(\Gamma_1^\dagger \Gamma_1 - \Gamma_o^\dagger \Gamma_o + 1) \tag{7.14}$$

The form of the internal product associated with a negative metric has been discussed in (3.28). The negative metric allows the cancellation of the spurious contributions. For instance, the fluctuation of the angular momentum in the deformed state disappears (as it should according to eq. (6.24))

$$< (\hat{J}^{(1)})^2 >= (\frac{\Im\omega}{2}) < ([\Gamma_1, \Gamma_1^\dagger] + [\Gamma_o, \Gamma_o^\dagger]) >= 0 \tag{7.15}$$

Another application of the metric (7.13) is the following calculation of the moment of inertia: in the BRST hamiltonian (6.21), the rotational-intrinsic coupling is given by the term $\hat{I}\hat{\Omega}$. Using the transformation (7.10) also in this case, we obtain in perturbation theory

$$\Delta E(I) = -I^2 < \hat{\Omega}^2 > /\omega = I^2/2\Im \tag{7.16}$$

where the negative metric allows to obtain a positive second order contribution as befits the concept of inertia.

iii) The quadratic ghost terms

$$\hat{H}_{gh}^{(2)} = i\pi\bar{\pi} - i\omega^2\bar{\eta}\eta \tag{7.17}$$

The ghosts are uncoupled through a transformation similar to (4.34)

$$\begin{aligned} a &= \frac{1}{\sqrt{2\omega}}\hat{\pi} - i\sqrt{\frac{\omega}{2}}\hat{\eta}; \quad b = \sqrt{\frac{1}{2\omega}}\hat{\bar{\pi}} + i\sqrt{\frac{\omega}{2}}\hat{\eta} \\ \bar{a} &= i\sqrt{\frac{1}{2\omega}}\hat{\bar{\pi}} + \sqrt{\frac{\omega}{2}}\hat{\bar{\eta}}; \quad \bar{b} = -i\frac{1}{\sqrt{2\omega}}\hat{\pi} + \sqrt{\frac{\omega}{2}}\hat{\bar{\eta}} \end{aligned} \tag{7.18}$$

where

$$1 = [a, \bar{a}]_+ = [b, \bar{b}]_+ \tag{7.19}$$

while all other anticommutators vanish. In terms of these operators $\hat{H}^{(2)}_{gh}$ becomes

$$\hat{H}^{(2)}_{gh} = \omega(\bar{a}a + \bar{b}b - 1) \tag{7.20}$$

Notice: i) the cancellation of the spurious constant ω in eq. (7.14) with the similar term in eq. (7.20). Thus (to leading order in x_o^{-1}) the ground state energy, which is a physical magnitude, is independent of ω (as it should); ii) the existing supersymmetry in the spectrum of spurious bosons and ghosts; and iii) the similarity between the spurious hamiltonian $\hat{H}^{(2)}_{sp} + \hat{H}^{(2)}_{gh}$ and the scaling operator in eq. (4.49).

In terms of the ghost operators defined in eqs. (7.18) the BRST charge $\hat{Q}$ reads:

$$\begin{aligned}\hat{Q} = \quad & [\sqrt{\frac{\omega}{2}}\hat{P} + \frac{i}{\sqrt{2\omega}}(I - \hat{J})]a \\ + \quad & [i\sqrt{\frac{\omega}{2}}\hat{P} + \frac{1}{\sqrt{2\omega}}(I - \hat{J})]\bar{b}\end{aligned} \tag{7.21}$$

whose leading order term $\hat{Q}^{(2)}$ is equal to:

$$\hat{Q}^{(2)} = -[i(\Gamma_1^\dagger + \Gamma_o^\dagger)a + (\Gamma_1 + \Gamma_o)\bar{b}] \tag{7.22}$$

States without $n = 0, 1$ phonons and also without a, b ghosts satisfy

$$a\,|\Psi> = b\,|\Psi> = \Gamma_1\,|\Psi> = \Gamma_o\,|\Psi> = 0 \tag{7.23}$$

Therefore they are physical to leading order, since

$$\hat{Q}^{(2)}|\Psi>= 0 \tag{7.24}$$

and consequently perturbative expansions may be constructed starting from such states.

7.1.3 The perturbative expansion

We proceed to calculate perturbative correction to the energies of vacuum and one-phonon states [31]. Since H_{BRST} has non zero frequency modes, it is free of the problems associated with broken symmetries. In the present calculation we assume no other term than (7.4) in the gauge fixing operator ($\hat{G} = \hat{G}^{(1)}$)

The expansion yields the residual hamiltonian H_{BRST}

$$\hat{H}_{res} = \hat{H}^{(2')} + \hat{H}^{(3)} + \hat{H}^{(4)} + ... \tag{7.25}$$

[31] The perturbative treatment in the product space of collective and intrinsic coordinates requires some special considerations which we discuss in section 9. Since in this abelian case we can work within an I-subspace, I is a constant and these considerations can be disregarded

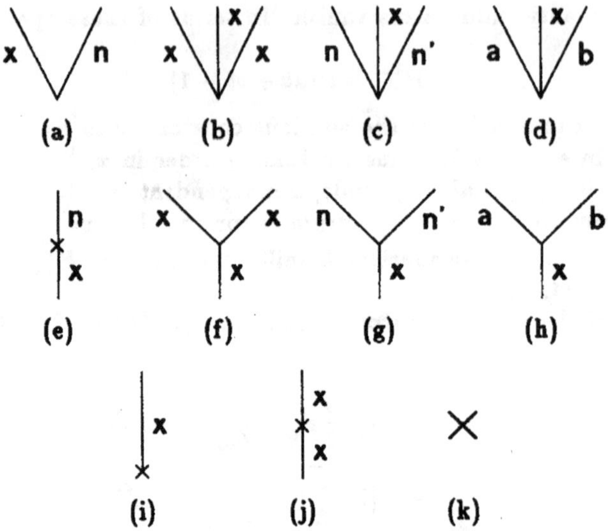

Figure 3: The vertices entering in the corrections (7.27) and (7.28). The single cross represents the ground state expectation value.

(a), (e)	$-\frac{I}{x_0^2(1+\delta_{n1})}\sqrt{\omega}\left(1-s\frac{\delta_{n1}}{\omega}\right)$
(b), (f)	$\frac{3}{2x_0\sqrt{2}}$
(c), (g)	$-\frac{\omega}{2x_0\sqrt{2}}\left[\delta_{nn'}\left(\frac{\delta_{n1}}{\omega^3}+2\delta_{n0}\right)+\left(1-\frac{n}{\omega}\right)(1-\delta_{nn'})\right]$
(d), (h)	$-\frac{i\omega}{2x_0\sqrt{2}}$
(i)	$\frac{1}{4x_0\sqrt{2}}\left(3+\frac{1}{\omega}\right)$
(j)	$\frac{1}{8x_0^2}\left(3+\frac{1}{\omega}\right)$
(k)	$\frac{1}{32x_0^2}\left(3+\frac{2}{\omega}+\frac{3}{\omega^2}\right)+\frac{I^2}{4x_0^4}\left(3+\frac{1}{\omega}\right)$

Table 2: Value of the vertices of the residual hamiltonian. We take $s=1$ for the backward vertices $(a),(b),(c)$ and (d) and $s=-1$ for the forward vertices $(e),(f),(g)$ and (h) in fig.3 .

where

$$
\begin{aligned}
\hat{H}^{(2')} &= -\frac{I}{x_o}\hat{\Omega}\hat{q}_1 - \frac{I}{x_o^2}(\hat{q}_1\hat{p}_2 - \hat{q}_2\hat{p}_1) + \frac{I^2}{x_o^4}(1 + 3\hat{q}_1^2 + \hat{q}_2^2) \\
\hat{H}^{(3)} &= \frac{1}{2x_o}\hat{q}_1(\hat{q}_1^2 + \hat{q}_2^2) - \hat{\Omega}(\hat{q}_1\hat{p}_2 - \hat{q}_2\hat{p}_1) \\
&\quad - \frac{\omega}{2x_o}\hat{q}_1(1 - \bar{a}a - \bar{b}b + i\bar{a}\bar{b} + iba) \\
\hat{H}^{(4)} &= \frac{1}{8x_o^2}(\hat{q}_1^2 + \hat{q}_2^2)^2
\end{aligned}
\tag{7.26}
$$

This expansion gives rise to the vertices shown in fig. 3. Their values are displayed in table 2.

The diagrammatic corrections of $O(x_o^{-2})$ to the energies of the ground state and of the one-phonon state and of order $O(x_o^{-4})$ to the I-dependent part of these energies are given in fig. 4. The value of the corresponding corrections is displayed in tables 3 and 4, respectively. Although the spurious frequency ω appears in each diagram, it cancels from the final (physical) sums.

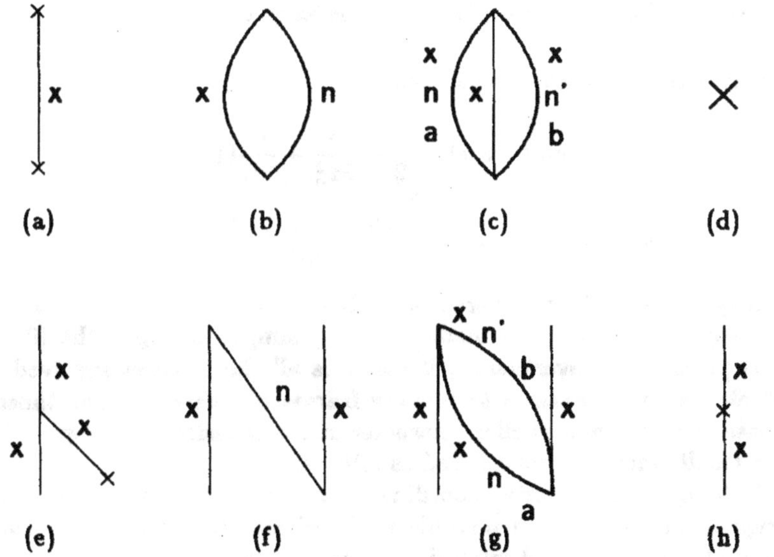

Figure 4: Diagrammatic contributions to the energy of the ground state (a) - (d) and to the energy of the one-phonon state (e) - (h).

(a)	$-\frac{9}{32x_o^2}-\frac{3}{16x_o^2\omega}-\frac{1}{32x_o^2\omega^2}$
(b)	$\frac{3}{4}\frac{I^2}{x_o^4}-\frac{I^2}{4x_o^4\omega}$
(c)	$-\frac{3}{16x_o^2}+\frac{1}{8x_o^2\omega}-\frac{1}{16x_o^2\omega^2}$
(d)	$\frac{3}{32x_o^2}+\frac{1}{16x_o^2\omega}+\frac{3}{32x_o^2\omega^2}+\frac{3}{4}\frac{I^2}{x_o^4}+\frac{I^2}{4x_o^4\omega}$

Table 3: Value of the diagrammatic contributions to the energy of the ground state using the value of the vertices as in table 2 .

(e)	$-\frac{9}{8x_o^2}-\frac{3}{8x_o^2\omega}$
(f)	$\frac{3}{2}\frac{I^2}{x_o^4}$
(g)	$-\frac{3}{4x_o^2}+\frac{1}{4x_o^2\omega}$
(h)	$\frac{3}{8x_o^2}+\frac{1}{8x_o^2\omega}+\frac{3}{2}\frac{I^2}{x_o^4}$

Table 4: Value of the diagrammatic contributions to the energy of the one-phonon state using the value of the vertices as in table 2 .

Finally, the perturbed energies read[32]

$$E(\text{ground state}) = \frac{1}{2} - \frac{3}{8x_o^2} + \frac{I^2}{2\Im}(1+3/x_o^2) \tag{7.27}$$

$$E(\text{one-phonon state}) = \frac{3}{2} - \frac{15}{8x_o^2} + \frac{I^2}{2\Im}(1+9/x_o^2) \tag{7.28}$$

which may be verified using perturbatively eq. (6.9).

The application has been made to a very simple example, the (0+1) Mexican hat potential. Nevertheless, it contains all the features involved in more complicated transformations to moving frames of reference (but those effects associated with the non abelian character of the transformation group).

Our results may be summarized as follows:

i) Although the collective coordinates are treated on the same footing as the original ones, it is both possible and desirable to factorize the total wave function into collective and intrinsic subspaces.

[32]This result is far from trivial. In fact, there appear to exist problems associated with the use of the gauge fixing operator $\hat{G}^{(1)}$ in the exponentiation of the Faddeev-Popov determinant if ghosts are not used (see appendix 7.A). The need to include properly the contribution of the spurious sector in perturbation theory is emphasized in appendix 7.B

ii) The set of intrinsic variables is made from the original variables of the problem, the Lagrange multipliers and the ghosts. The determination of the basis in the intrinsic subspace is made through a "classical" minimization yielding a "deformed " solution and through the elementary normal modes around this minimum. This procedure yields the inertial parameter of the collective motion, the intrinsic excitations and, in addition, two spurious modes associated with the well known zero frequency mode in deformed systems and with the Lagrange multiplier. The separation between these spurious modes results in two modes with the same (arbitrary) frequency, one of them having a negative metric.

iii) We use an expansion in terms of the inverse of the (adimensional) deformation parameter. The quadratic expressions obtained in ii) contain all the leading terms of this expansion. The effects associated with both the residual interactions and with a completely quantal treatment are treated perturbatively in terms of an (asymptotic) series in powers of the expansion parameter. We have performed explicitly the perturbative calculation for the ground state energy, the frequency of the real phonon and the moment of inertia in the two bands.

Although in general the ghosts play an essential role in the BRST calculations, in this abelian example they could have been uncoupled through the use of the exact conjugate angle $\hat{\theta}$ as gauge fixing operator $\hat{G}$

$$\hat{\theta} = tg^{-1}(\frac{\hat{q}_2}{x_o + \hat{q}_1}) \tag{7.29}$$

The vertices (d) and (h) in fig. 3 would be absent in this case[33].

Finally we note again the similarity with electromagnetism (cf. table 1). In that case we have Lagrange multipliers $A_o(x_i)$ (which play the same role as Ω in our case) imposing the constraint $\nabla.\vec{E}(x_i) = 0$ (the Gauss' law). The canonical conjugates $E_o(x_i)$ of the $A_o(x_i)$ vanish (which corresponds to the condition $P = 0$ in the present case). Due to negative metric the $n = 0$ phonon behaves as the four component of the vector potential in the Bleuler-Gupta formalism [22].

7.2 The BCS description

We now illustrate the application of the method to the case of a many-body problem. We consider a system of fermions which admits a BCS description. In this case, the relevant symmetry operation is the (abelian) transformation

$$\hat{T} = exp[-i\phi\hat{n}] \tag{7.30}$$

where $\hat{n}$ is the operator corresponding to the number of pairs of particles and plays the same role as the particle angular momentum $\hat{J}$ did in the previous example.

[33]The reader is invited to rederive the value of the other vertices by expanding θ in powers of x_o^{-1} and to verify that the results (7.27) and (7.28) are reproduced in this case.

$$\hat{n} = \sum_k \hat{n}_k; \qquad \hat{n}_k = \frac{1}{2}(a_k^\dagger a_k + a_{\bar{k}}^\dagger a_{\bar{k}}) \tag{7.31}$$

The operator $a_{\bar{k}}^\dagger$ creates a particle in the time-reversed state of $a_k^\dagger$. Thus the summation in (7.31) extends to over half the single-particle spectrum.

The hamiltonian consists of a single-particle term plus a pairing interaction

$$\hat{H} = \sum_k 2\epsilon_k \hat{n}_k - V\hat{S}^\dagger \hat{S} \tag{7.32}$$

where ϵ_k is a single-particle energy, V the strength of the pairing interaction and $\hat{S}^\dagger$, the transfer operator,

$$\hat{S}^\dagger = \sum_k \hat{s}_k^\dagger; \qquad \hat{s}_k^\dagger = a_k^\dagger a_{\bar{k}}^\dagger \tag{7.33}$$

The number operators $\hat{n}_k$ and the hamiltonian are invariant with respect to the transformation (7.30), while the particle and the transfer operators transform as

$$\begin{aligned} a_k^\dagger &\rightarrow exp[-i\frac{1}{2}\phi]\, a_k^\dagger \\ \hat{s}_k^\dagger &\rightarrow exp[-i\phi]\, \hat{s}_k^\dagger \end{aligned} \tag{7.34}$$

The BCS eigenstates do not carry irreducible representations of the group of transformations (7.30). On the other hand, the solutions of the corresponding BRST hamiltonian (6.21) have a good macroscopic symmetry, namely

$$\Psi_N = \frac{1}{\sqrt{2\pi}} exp[iN\phi]\; | >_N \tag{7.35}$$

The "rotational" ground band is the sequence of ground states of systems with an even number of fermions [34]. The calculation can again be restricted to a subspace labelled by the eigenvalues of the collective operator $\hat{N}$ and proceeds along the same steps as in the previous case, namely i) the variational procedure determining the ground state, ii) the determination of the elementary modes of excitation and iii) the perturbative calculation.

7.2.1 The variational procedure

The minimization eqs. (4.19) yield, in this abelian case,

$$< |\hat{P}| >_N \; = \; < |\hat{G}| >_N = 0 \tag{7.36}$$

$$< |\hat{n}| >_N \; = \; N \tag{7.37}$$

The "intrinsic" states $| >_N$ are determined through a minimization procedure:

$$\delta < |(\hat{H} - \Omega\hat{n})| >_N = 0 \tag{7.38}$$

where the expectation value of the Lagrange multiplier,

$$< |\hat{\Omega}| >_N = \Omega, \tag{7.39}$$

is determined by the condition that (7.37) holds true. The invariance (7.38) yields

$$\Omega = \frac{\partial}{\partial N} < |\hat{H}| >_N = N/\Im^{(1)} \tag{7.40}$$

where the ratio N/Ω defines the "moment of inertia" $\Im^{(1)}$. Therefore, the expectation value of the hamiltonian contains a rotational energy.

$$< |\hat{H}| >_N = const. + \frac{1}{2\Im^{(1)}} N^2 \tag{7.41}$$

Since the number of pairs of particles is not conserved, the expectation value of the transfer operator $\hat{S}^\dagger$ does not vanish. This expectation value is labelled Δ/V (cf. eq.(7.45)) and plays the same role as the expectation value x_o of the coordinate $z = q_1 + iq_2$ in the previous example. In fact, the gauge conditions in both cases amount to require that the order parameter is a real number. We thus define the quadratic fermion hamiltonian:

$$\begin{aligned} \hat{H}_{BCS} &= \frac{\Delta^2}{V} + \Omega N + 2(\epsilon_k - \Omega)\hat{n}_k - \Delta(\hat{S}^\dagger + \hat{S}) \\ &= < |\hat{H}| >_N + \sum_k E_k \hat{\nu}_k \end{aligned} \tag{7.42}$$

In this equation, the BCS hamiltonian has been diagonalized by means of a Bogoliubov-Valatin transformation to quasi-particles

$$\alpha_k^\dagger = U_k a_k^\dagger - V_k a_{\bar{k}} \qquad U_k^2 + V_k^2 = 1 \tag{7.43}$$

We define

$$\begin{aligned} E_k &\equiv [(\epsilon_k - \Omega)^2 + \Delta^2]^{1/2} \\ \hat{\nu}_k &\equiv \alpha_k^\dagger \alpha_k + \alpha_{\bar{k}}^\dagger \alpha_{\bar{k}} \\ \gamma_k^\dagger &\equiv \alpha_k^\dagger \alpha_{\bar{k}}^\dagger \end{aligned} \tag{7.44}$$

The minimization condition is equivalent to the requirement of self-consistency

$$\frac{\Delta}{V} \equiv < |\hat{S}^\dagger| >_N = \sum_k U_k V_k \tag{7.45}$$

which must be satisfied in addition to eq. (7.37).

In the new representation the transfer and the number operators read

$$\begin{aligned}\hat{S}^{\dagger} &= \frac{\Delta}{V}+\sum_{k}(U_k^2\gamma_k^{\dagger}-V_k^2\gamma_k-U_kV_k\hat{\nu}_k)\\ \hat{n} &= \sum_{k}(V_k^2+U_kV_k(\gamma_k^{\dagger}+\gamma_k)+\frac{1}{2}(U_k^2-V_k^2)\hat{\nu}_k)\end{aligned} \tag{7.46}$$

These operators are used within the RPA in the next step of the calculation

7.2.2 The normal modes

In an even system the lower intrinsic excited states are two-quasi-particle states. We are specially interested in those states $\gamma_k^{\dagger}| >_N$ in which the quasi-particles are in time-reversed orbits. The RPA approximates $\gamma_k^{\dagger}$ by a boson creation operator and retains only linear terms in the one-body operators an quadratic terms in two-body operators (such as the hamiltonian)

$$\begin{aligned}\hat{S}_{RPA}^{\dagger} &= \sum_{k}(U_k^2\gamma_k^{\dagger}-V_k^2\gamma_k)\\ \hat{n}_{RPA} &= \sum_{k}U_kV_k(\gamma_k^{\dagger}+\gamma_k)\\ \hat{H}_{RPA} &= \sum_{k}2E_k\gamma_k^{\dagger}\gamma_k-V\hat{S}_{RPA}^{\dagger}\hat{S}_{RPA}\\ &= \frac{1}{2\Im^{(2)}}\hat{n}_{RPA}^2+\sum_{n>1}\omega_n(\Gamma_n^{\dagger}\Gamma_n+\frac{1}{2})\end{aligned} \tag{7.47}$$

The transformation of $\hat{H}_{RPA}$ to normal modes yields finite frequency modes ($n > 1$) which play a similar role to the (real) radial vibration in the case of the Mexican hat potential. In addition there is a zero frequency mode with no restoring force [35]. The moment of inertia $\Im^{(2)}$ and the RPA angle conjugate to $\hat{n}_{RPA}$ are obtained from the equations

$$[\hat{H}_{RPA},\hat{\theta}_{RPA}]=-\frac{i}{\Im^{(2)}}\hat{n}_{RPA} \tag{7.48}$$

$$[\hat{\theta}_{RPA},\hat{n}_{RPA}]=i \tag{7.49}$$

which yield

$$\hat{\theta}_{RPA} = i\sum_{k}\theta_k(\gamma_k^{\dagger}-\gamma_k)$$

$$\theta_k = -\frac{1}{4\Xi_2\Im^{(2)}}(\Xi_1\frac{\epsilon_k-\Omega}{E_k^2}+\Xi_2\frac{\Delta}{E_k^2})$$
$$\Im^{(2)} = \frac{\Delta}{4\Xi_2}(\Xi_1^2+\Xi_2^2)$$
$$\Xi_1 = \sum_k\frac{\epsilon_k-\Omega}{E_k^3};\qquad \Xi_2=\Delta\sum_k\frac{1}{E_k^3} \tag{7.50}$$

Since both $\hat{n}_{RPA}$ and $\hat{\theta}_{RPA}$ are linear in the boson variables they play the same role as $\hat{J}^{(1)}$ and $\hat{G}^{(1)}$ in the previous example. In fact, we can also take here $\hat{G}=\hat{\theta}_{RPA}$ or add extra terms as in eq. (7.29).

The important feature is that the spurious quadratic term in the BRST hamiltonian has the same form as in eq. (7.9). The transformation (7.10) to normal modes and the subsequent discussion concerning the negative metric of the $n=o$ phonon and the ghost quadratic hamiltonian applies as well. We finally obtain the quadratic hamiltonian

$$\hat{H}^{(2)}_{BRST} = const. + \frac{1}{2\Im^{(1)}}\hat{N}^2 + \sum_{n>1}\omega_n(\Gamma_n^\dagger\Gamma_n+\frac{1}{2})$$
$$+\omega(\Gamma_1^\dagger\Gamma_1-\Gamma_o^\dagger\Gamma_o+\bar{a}a+\bar{b}b) \tag{7.51}$$

The problem treated in the present subsection is much more complicated than the problem of a particle moving in a Mexican hat potential. This complexity manifest itself in the appearance of (many) real bosons ω_n $(n>1)$. However, the treatment of the rotational energy and of the spurious degrees of freedom is as simple as in the previous case.

States with any number of real bosons but in the vacuum state (7.23) for the spurious degrees of freedom provide an adequate basis for perturbation procedures since they are physical in the BRST sense.

7.2.3 The perturbative calculations

Any available method to improve systematically over the RPA (like bosonic expansions [32] or the nuclear field theory [33]) may be used here. A common ingredient to all these methods is the replacement of the operators $\gamma_k^\dagger, \gamma_k$ by their expressions in terms of the orthonormal modes, among which we must include the phonons $n=o,1$. The forward (λ_{nk}) and backward (μ_{nk}) amplitudes of the real phonons $(n>1)$ are obtained within the usual RPA formalism. We write

$$\gamma_k^\dagger \rightarrow \sum\nolimits_{n>1}(\lambda_{nk}\Gamma_n^\dagger+\mu_{nk}\Gamma_n)-\theta_k\hat{n}_{RPA}+iU_kV_k\hat{\theta}_{RPA}$$
$$= \sum\nolimits_{n\geq o}(\lambda_{nk}\Gamma_n^\dagger+\mu_{nk}\Gamma_n) \tag{7.52}$$

where

$$\lambda_{n1} = (\frac{U_kV_k}{\Im^{(2)}\omega} - \theta_k)\sqrt{\frac{1}{2}\Im^{(2)}\omega} \quad \lambda_{no} = -\theta_k\sqrt{\frac{1}{2}\Im^{(2)}\omega}$$

$$\mu_{n1} = -(\frac{U_kV_k}{\Im^{(2)}\omega} + \theta_k)\sqrt{\frac{1}{2}\Im^{(2)}\omega} \quad \mu_{no} = -\theta_k\sqrt{\frac{1}{2}\Im^{(2)}\omega} \qquad (7.53)$$

We have used the fact that $\hat{n}_{RPA}$ and $\hat{\theta}_{RPA}$ are orthogonal to the real modes and the transformation (7.10) between the spurious modes. This replacement yields, for instance, the vertices of the residual interaction. Once these are obtained, the only complication relative to usual perturbation theory is the existence of the additional phonon $n = o$ with a negative metric, which is very easy to take into account.

The perturbative calculations require an expansion parameter in order to group together the contributions of the same order in this parameter, since the cancellation of the arbitrary constant ω takes place order by order. This expansion parameter cannot be the order of perturbation theory: in fig. 4 both first order diagrams (d),(h) and second order diagrams contribute to the same order $(O(x_o^{-2}))$ in this cancellation.

The order parameter $\frac{\Delta}{V}$ is also a large quantity, since it involves a summation of positive terms over all states that enter significantly in the construction of the BCS solution (eq. (7.45)). Therefore the choice of $\frac{V}{\Delta}$ as an expansion parameter is based on the existence of a large effective degeneracy among the single-particle states. This choice is consistent with the approximations involved in the RPA, which privilege the two-quasi-particle creation and annihilation terms in one-body operators. For instance, the expectation value of the square of $\hat{n}_{RPA}$ is of the same order as the number of terms entering in the second eq. (7.47) $(O(\frac{\Delta}{V}))$. Thus the operator $\hat{n}_{RPA}$ is of $O(\sqrt{\frac{\Delta}{V}})$ which is large relative to the matrix elements of the terms conserving the number of quasi-particles $(O(1))$. The operator $\hat{\theta}_{RPA}$ is of $O(\sqrt{\frac{V}{\Delta}})$ according to eq. (7.46), while the remaining terms in $\hat{\theta}$ are at most of $O(\frac{V}{\Delta})$. If we assume that the hamiltonian H_{RPA} is of $O(1)$, the moment of inertia is of $O(\frac{\Delta}{V})$ (cf. eq. (7.48)). Thus $\sqrt{\frac{V}{\Delta}}$ plays the same role as x_o^{-1} did in the simple example discussed in subsection 7.1.

This evaluation may be readily extended to other one- and two-body operators and, in particular, to the different terms in the total hamiltonian. We thus obtain a classification of the vertices of the residual interaction in powers of $\sqrt{\frac{V}{\Delta}}$ analogous to the one appearing in table 2. In the assignment of a definite order to each diagrammatic contribution the number of fermion loops must also be considered . A reasonable procedure is to assume that all single-particle levels are degenerate (in which case each diagram is proportional to a definite power of this degeneracy) and to group together all diagrams contributing to the same power.

In the second of refs. [9], perturbative calculations have been carried out for the case of particles in two degenerate levels and coupled through a pairing interaction. The hamiltonian treated there is identical to the case of H_{BRST} in which the ghost decouple because the exact angle is used (eq. 7.29).

The one-level case was solved in the first of refs. [8] using the formalism associated with the Coulomb gauge in section 5.

appendix 7.A The calculation in the Coulomb gauge

In this appendix we treat the Mexican hat potential without using ghosts. This can be achieved, for instance, by using the exponentiation (5.21) of the Faddeev-Popov determinant and the Coulomb gauge, as presented in subsection 5.2. The result displays limitations on the choice of the gauge-fixing function which are not present in the previous calculation of subsubsection 7.1.3 (cf. the footnote on p.75).

For simplicity we work in the subspace $I = 0$. The hamiltonian (5.22) reads[34]

$$\hat{H}_{eff} = \lim_{D\to 0}\left[\frac{\hat{p}_1^2}{2} + \frac{\hat{p}_2^2}{2} + V(\hat{q}_1, \hat{q}_2) + \frac{\hat{J}^2}{2Dx_o^2} + \frac{x_o^2\hat{G}^2}{2A} + i\delta(0)ln(1+\hat{g})\right] \quad (7.54)$$

where

$$[\hat{G}, \hat{J}] \equiv i(1+\hat{g}) \quad (7.55)$$

We have replaced the parameters A and D in eq. (5.22) by A/x_o^2 and Dx_o^2, respectively, in order to homogenize the quadratic hamiltonian with respect to the order parameter x_o. The coordinate q_1 is again measured from x_o. The angular momentum $\hat{J}$ and the linear terms $\hat{J}^{(1)}$ and $\hat{G}^{(1)}$ are the same as in eqs. (7.2)-(7.4).

The quadratic hamiltonian is

$$\begin{aligned} \hat{H}^{(2)}_{eff} &= \frac{1}{2}(\hat{p}_1^2 + \hat{q}_1^2) + \frac{1+D}{2D}\hat{p}_2^2 + \frac{1}{2A}\hat{q}_2^2 \\ &= \Gamma_x^\dagger\Gamma_x + \frac{1}{2} + \omega(\Gamma_y^\dagger\Gamma_y + \frac{1}{2}) \end{aligned} \quad (7.56)$$

where $\Gamma_x^\dagger$ is defined in eq. (7.8) and

$$\begin{aligned} \omega &\equiv \sqrt{\frac{1+D}{AD}} \\ \Gamma_y^\dagger &\equiv \sqrt{\frac{1}{2A\omega}}\hat{q}_2 - i\sqrt{\frac{A\omega}{2}}\hat{p}_2 \end{aligned} \quad (7.57)$$

The spurious boson is eliminated in the limit $D \to 0$, as discussed in subsection 5.2. However, this limit can only be taken in final physical expressions. The calculations must be carried out using finite (arbitrary) values of A and D.

According to (7.56), the treatment of the spurious degree of freedom in the Coulomb gauge appears to be much simpler than in the Lorentz gauge (involving Lagrange multipliers and ghosts). However, as usual in field theory,

[34] Although the constant $\delta(0)$ has been evaluated in appendix 5.A, we keep the original expression for the time being.

the calculations in the Coulomb gauge are indeed more cumbersome. Here this fact is due to the D-dependence of the residual hamiltonian, which reads

$$
\begin{aligned}
\hat{H}^{(3)}_{eff} &= \frac{1}{2x_o}\hat{q}_1(\hat{q}_1^2+\hat{q}_2^2) + \frac{1}{2Dx_o}(2\hat{q}_1\hat{p}_2 - \hat{p}_1[\hat{p}_2,\hat{q}_2]_+) \\
&+ \frac{x_o}{2A}[\hat{q}_2,\hat{G}^{(2)}]_+ + i\delta(0)\hat{g}^{(1)} \\
\hat{H}^{(4)}_{eff} &= \frac{1}{8x_o^2}(\hat{q}_1^2+\hat{q}_2^2)^2 + \frac{1}{D}(\hat{q}_1\hat{p}_2 - \hat{q}_2\hat{p}_1)^2 \\
&+ \frac{x_o}{2A}(x_o(\hat{G}^{(2)})^2 + [\hat{q}_2,\hat{G}^{(3)}]_+) + i\delta(0)\hat{g}^{(2)}
\end{aligned} \tag{7.58}
$$

In the first place we perform the calculation by assuming that the gauge-fixing function is an odd function of the exact angle $\hat{\theta}$ (7.29). In such a case

$$
\begin{aligned}
\hat{G} &= \frac{\hat{q}_2}{x_o+\hat{q}_1} + \alpha\left(\frac{\hat{q}_2}{x_o+\hat{q}_1}\right)^3 + \ldots \\
\hat{G}^{(2)} &= -\frac{1}{x_o^2}\hat{q}_1\hat{q}_2 \\
\hat{G}^{(3)} &= \frac{1}{x_o^3}(\hat{q}_1^2+\alpha\hat{q}_2^2) \\
\hat{g}^{(1)} &= 0 \\
\hat{g}^{(2)} &= \frac{1+3\alpha}{x_o^2}\hat{q}_2^2
\end{aligned} \tag{7.59}
$$

where α depends on the particular choice of the function[35].

The calculation of the ground state energy only requires the vertices (b), (c), (i) and (k) in fig.3. In the vertex (c) we take $n = n' = y$. In particular, the contribution to the expectation value (k) of the terms containing the gauge-fixing function yields

$$
\begin{aligned}
\frac{x_o^2}{2A} < \hat{G}^2 > + i\delta(0) < \hat{g}^{(2)} > &= \frac{3\omega}{8x_o^2} + \frac{1+D}{2x_o^2 D\omega}[i\delta(0) + 3\alpha(\frac{\omega}{2} + i\delta(0))] \\
&= \frac{3\omega}{8x_o^2} - \frac{\omega}{4x_o^2}(\frac{1+D}{\omega D})
\end{aligned} \tag{7.60}
$$

Therefore, the result is independent (as it should) of the particular choice of the function of the angle (i.e., of the value of α), provided that the value of $-\omega/2$ is assigned to $i\delta(0)$. The value derived in appendix 5.A has thus been verified. It has been used on the second line of eq. (7.60).

The final value of the vertices is given in table 5. The diagrammatic contributions to the ground state energy are represented by the graphs (a), (c) and (d) on fig. 3, which are evaluated in table 6.

[35] For instance, the exact angle θ corresponds to $\alpha = -1/3$.

(b)	$\frac{3}{2x_o\sqrt{2}}$
(c)	$\frac{1}{Dx_o\sqrt{2}}\left[1-2\omega D+\frac{1}{2\omega}+\omega D^2+\frac{D}{2\omega}+O(D^2)\right]$
(i)	$\frac{1}{4x_o\omega D\sqrt{2}}\left[1+3\omega D+D-2D^2\omega^2+O(D^{3/2})\right]$
(k)	$\frac{1}{8x_o^2D^2\omega}\left[1+\frac{3}{4\omega}-4D\omega+\frac{3D}{2}+4D^2\omega^2-\frac{5D^2\omega}{4}+\frac{3D}{2\omega}+O(D^{1/2})\right]$

Table 5: Value of the vertices of the residual hamiltonian (7.58) using the gauges (7.59).

(a)	$-\frac{1}{32x_o^2D^2\omega^2}-\frac{3}{16x_o^2D\omega}-\frac{5}{32x_o^2}-\frac{1}{16x_o^2D\omega^2}$
(c)	$-\frac{1}{8x_o^2D^2\omega}-\frac{1}{16x_o^2D^2\omega^2}+\frac{1}{2x_o^2D}-\frac{\omega}{2x_o^2}-\frac{1}{16x_o^2}-\frac{1}{8x_o^2D\omega^2}$
(d)	$\frac{1}{8x_o^2D^2\omega}+\frac{3}{32x_o^2D^2\omega^2}-\frac{1}{2x_o^2D}+\frac{3}{16x_o^2D\omega}+\frac{\omega}{2x_o^2}-\frac{5}{32x_o^2}+\frac{3}{16x_o^2D\omega^2}$

Table 6: Value of the diagrammatic contributions to the energy of the ground state using the value of the vertices as in table 5 .

In the limit $D \to 0$, the final correction of $O(x_o^{-2})$ to the ground state energy is

$$\Delta E(g.s., I=0) = \frac{3}{8x_o^2} \tag{7.61}$$

which agrees with (7.27). In particular, it is finite (the partial contributions containing negative powers of D have cancelled out) and independent of A. Therefore we have verified: i) that the calculation with the formalism developed for the Coulomb gauge also yields the correct value and ii) that the result is independent of the selected gauge, provided it is an odd function of the angle θ.

We repeat now the calculation using the (linear) gauge, as in subsubsection 7.1.3

$$\hat{G} = \frac{1}{x_o}\hat{q}_2 \tag{7.62}$$

The new value of the vertices and of the diagrammatic corrections are given in tables 7 and 8, respectively. Unlike with the previous choice (7.59), the result diverges in the limit $D \to 0$ and the finite term depends on the value of A. No change on the value of $i\delta(0)$ can alter this negative conclusion about the validity of the choice (7.62). Cf. also p. 49. Note, however, that the calculation in the Lorentz gauge with ghosts using (7.62) yields exact results.

(b)	$\frac{3}{2x_o\sqrt{2}}$
(c)	$\frac{1}{Dx_o\sqrt{2}}\left[1-\omega D+\frac{1}{2\omega}+\omega D^2+\frac{D}{2\omega}+O(D^2)\right]$
(i)	$\frac{1}{4x_o\omega D\sqrt{2}}\left[1+2D\omega^2+3\omega D+D-2D^2\omega^2+O(D^{3/2})\right]$
(k)	$\frac{1}{8x_o^2D^2\omega}\left[1+\frac{3}{4\omega}-2D\omega+\frac{3D}{2}+2D^2\omega^2+\frac{3D^2\omega}{4}+\frac{3D}{2\omega}+O(D^{1/2})\right]$

Table 7: Value of the vertices of the residual hamiltonian (7.58) using the gauge (7.62).

(a)	$-\frac{1}{32x_o^2D^2\omega^2}-\frac{\omega^2}{8x_o^2}-\frac{1}{8x_0^2D}-\frac{3}{16x_o^2D\omega}-\frac{3\omega}{8x_o^2}-\frac{9}{32x_o^2}-\frac{1}{16x_o^2D\omega^2}$
(c)	$-\frac{1}{8x_o^2D^2\omega}-\frac{1}{16x_o^2D^2\omega^2}+\frac{1}{4x_o^2D}-\frac{\omega}{8x_o^2}-\frac{5}{32x_o^2}-\frac{1}{8x_o^2D\omega^2}$
(d)	$\frac{1}{8x_o^2D^2\omega}+\frac{3}{32x_o^2D^2\omega^2}-\frac{1}{4x_o^2D}+\frac{3}{16x_o^2D\omega}+\frac{\omega}{4x_o^2}+\frac{3}{32x_o^2}+\frac{3}{16x_o^2D\omega^2}$

Table 8: Value of the diagrammatic contributions to the energy of the ground state using the value of the vertices as in table 7 .

appendix 7.B Perturbative calculations with the original hamiltonian H in the deformed basis

The original hamiltonian H in the deformed basis yields

$$\begin{aligned} H &= \frac{1}{2}(p_1^2+q_1^2) + \frac{1}{2}(p_2^2+\omega^2 q_2^2) \\ &\quad + \frac{1}{2x_o}q_1(q_1^2+q_2^2) + \frac{1}{8x_o^2}(q_1^2+q_2^2)^2 \end{aligned} \tag{7.63}$$

The breakdown of symmetry manifests itself by the presence of a zero-frequency boson. We have added a restoring force in the q_2-direction and we will let $\omega \to 0$ in the final result. The value of the vertices in fig. 3 is given in table 9.

(b), (f)	$\frac{3}{2\sqrt{2}x_o}$
(c), (g)	$\frac{1}{2\sqrt{2}x_o\omega}$
(i)	$\frac{1}{4\sqrt{2}x_o}(3+\frac{1}{\omega})$
(j)	$\frac{1}{8x_o^2}(3+\frac{1}{\omega})$
(k)	$\frac{1}{32x_o^2}(3+\frac{2}{\omega}+\frac{3}{\omega^2})$

Table 9: The vertices of the residual interaction of the original hamiltonian H .

The corrections to the energies of the ground state and of the one-phonon state are

$$\Delta E(\text{ground state}) = -\frac{1}{2x_o^2}$$

$$\Delta E(\text{one-phonon}) = \frac{1}{16x_o^2}\left(\frac{1}{\omega^2} - \frac{2}{\omega} - 17\right) \tag{7.64}$$

Although the vertices diverge as $\omega \to 0$, the ground state correction is finite, albeit wrong (cf. eqs. (7.27) and (7.28)). This is also the case for the finite term in the one-phonon case, which in addition diverges as $\frac{1}{\omega^2}$.

An alternative calculation disregards the zero-frequency boson. This amounts to put $\omega = \infty$ in table 9. The resulting corrections are:

$$\Delta E(\text{ground state}) = -\frac{1}{4x_o^2}$$

$$\Delta E(\text{one-phonon}) = -\frac{21}{16x_o^2} \tag{7.65}$$

which amount to approximately $\frac{2}{3}$ of the $O(x_o^{-2})$ (cf. eqs. (7.27) and (7.28)). Thus, in this simple example, 30 % of the corrections are linked to the spurious sector.

Therefore, we must apply the BRST procedure (or any other method that may properly take into account the spurious sector) as soon as we require an accuracy going beyond the level of independent modes of excitation.

8 The rotational transformations in three dimensions

The solution to the BRST hamiltonian discussed in section 7 for abelian cases can be readily extended to non-abelian transformations. In the present section we treat the case of rotations in three dimensions.

In subsection 8.1 we review some general features of these rotational descriptions, such as gauge-fixing procedures and residual symmetries. We start from the BRST hamiltonian defined in eq. (4.18), which can be applied directly to the case of a triaxial rotating system[36]. In subsection 8.2 we discuss the modifications to be introduced if there is only a partial breakdown of the symmetry (as for systems having a mean-field approximation with axial symmetry), while in subsection 8.3 we study systems with triaxially deformed mean-fields at very high angular momenta.

8.1 The general features of systems rotating in three dimensions

8.1.1 The hamiltonian and the collective representation

We construct the Hilbert space $\mathcal{H}_4$ in which the collective variables are the three Euler angles describing the position of a rotating frame. The BRST charge is given in eq. (4.5). We obtain the hamiltonian (4.18) by choosing the operator $\hat{\rho}$ as in (4.17).

The hamiltonian $\hat{H}_{BRST}$ is cyclic with respect to the collective angular variables. Thus we work within the I, M subspace (since I^2 and I_3' are constants of the motion) and we factorize the wave function as in the case of the unified models [2]

$$\Psi_{IMKn}(\phi_v, \Omega_v, \eta_v, \bar{\pi}_v, q_i) = \sqrt{\frac{2I+1}{8\pi^2}}\; D^I_{MK}(\phi_v)\; \chi_{In}(\Omega_v, \eta_v, \bar{\pi}_v, q_i) \tag{8.1}$$

Angular momentum projection is achieved as described in eq. (6.24). In the three-dimensional case,

$$\hat{J}_v^2 = \hat{I}_v^2 - [(\hat{J}_v + 2\hat{I}_v)\hat{\bar{\pi}}_v, \hat{Q}]_+ \tag{8.2}$$

so that $\hat{J}_v^2$ equals $\hat{I}_v^2$ up to a null operator. Therefore, if we construct the state Ψ_{IM} such that it is an eigenstate of $\hat{H}_{BRST}$ and of $\hat{I}^2$ and it is annihilated by $\hat{Q}$ from (4.14) and (8.2) we have that

$$\hat{J}_v^2 \Psi_{IM} = I(I+1)\Psi_{IM} + |null\ \ state > \tag{8.3}$$

[36]Throughout this section we use the Levi-Civita tensor ϵ_{vwn} as the group structure constant.

Since we know that a null state is irrelevant as far as overlaps are concerned, the state Ψ_{IM} is for all practical purposes an angular momentum projected state. This is so to the extent that Ψ_{IM} becomes more physical as the approximation is improved.

8.1.2 The residual symmetries

Consider the following operators:

$$\begin{aligned} \hat{N}_v &= \epsilon_{vws}\hat{\Omega}_w\hat{P}_s \\ \hat{\tau}_v &= -i\epsilon_{vws}\hat{\eta}_w\hat{\pi}_s \\ \hat{\bar{\tau}}_v &= -i\epsilon_{vws}\hat{\bar{\eta}}_w\hat{\bar{\pi}}_s \end{aligned} \tag{8.4}$$

We define

$$\hat{L}_v \equiv \hat{N}_v + \hat{\tau}_v + \hat{\bar{\tau}}_v + \hat{f}_v \tag{8.5}$$

The four operators in $\hat{L}_v$ satisfy an SU(2) algebra. The three operators (8.4) generate rotations of the vector operators $(\vec{\Omega}, \vec{P})$, $(\vec{\eta}, \vec{\pi})$ and $(\vec{\bar{\eta}}, \vec{\bar{\pi}})$, respectively. It is easy to verify that they are "null" operators:

$$\hat{N}_v + \hat{\bar{\tau}}_v = [\hat{Q}, i\epsilon_{vws}\hat{\bar{\eta}}_w\hat{\Omega}_s]_+ \tag{8.6}$$

$$\hat{\tau}_v + \hat{f}_v = [\hat{Q}, \hat{\bar{\pi}}_v]_+ \tag{8.7}$$

Consider further the following three 180-degree rotation operators

$$\hat{\mathcal{R}}_v = \hat{\mathcal{R}}_v^J(\hat{\mathcal{R}}_v^I)^{-1}\hat{\mathcal{R}}_v^\Omega\hat{\mathcal{R}}_v^{gh} \tag{8.8}$$

where

$$\begin{aligned} \hat{\mathcal{R}}_v^J &= \exp[i\pi\hat{J}_v] \\ \hat{\mathcal{R}}_v^I &= \exp[i\pi\hat{I}_v] \\ \hat{\mathcal{R}}_v^\Omega &= \exp[i\pi\hat{N}_v] \\ \hat{\mathcal{R}}_v^{gh} &= \exp[i\pi(\hat{\bar{\tau}}_v + \hat{\tau}_v)] \end{aligned} \tag{8.9}$$

We note that

$$\hat{\mathcal{R}}_v = 1 + \quad null\ operator \tag{8.10}$$

and, in consequence,

$$\hat{\mathcal{R}}_v|phys> = |phys> + |null\ \ state> \tag{8.11}$$

Also, from Jacobi's identity

$$[\hat{\mathcal{R}}_v, \hat{Q}] = 0 \tag{8.12}$$

In the case of pure quadrupole triaxial deformations it is both possible and desirable to choose operators $\hat{G}_v$ that satisfy[37]

$$\hat{\mathcal{R}}_v^J \hat{G}_w \hat{\mathcal{R}}_v^{J\dagger} = (2\delta_{vw} - 1)\hat{G}_w \tag{8.13}$$

From eq. (8.13) and with the choice (4.17) for $\hat{\rho}$, we can also check that

$$[\hat{\rho}, \hat{\mathcal{R}}_v] = 0 \tag{8.14}$$

and hence using Jacobi's identity

$$[\hat{H}_{BRST}, \hat{\mathcal{R}}_v] = 0 \tag{8.15}$$

These identities imply that physical wavefunctions can be simultaneously eigenvectors of $\hat{H}_{BRST}$ and $\hat{\mathcal{R}}_v$. But if

$$\hat{\mathcal{R}}_v |phys> = -|phys> \tag{8.16}$$

using (8.11)

$$-|phys> = |phys> + |null\ \ state> \tag{8.17}$$

thus $|phys>$ is a null state unless for all v

$$\hat{\mathcal{R}}_v |phys> = |phys> \tag{8.18}$$

Therefore the only eigenstates of interest in the triaxial case are those having eigenvalue one for the three rotation operators $\hat{\mathcal{R}}_v$ (i.e. the A-representation of the group D_2). This result is in agreement with the one in [2]. In the $BRST$ treatment we had to include also a 180-degree rotation of the angular velocity operators $\hat{\mathcal{R}}_v^{\Omega}$ and of the ghost operators $\hat{\mathcal{R}}_v^{gh}$ (8.9).

8.1.3 The construction of the basis

In section 7, the construction of the spurious sector of the intrinsic wave function χ_{In} approximately identifies the gauge fixing operator $\hat{G}$ with the conjugate variable to the generator $\hat{J}$. It also assumes that one of the normal modes of the original hamiltonian is associated with a rotational degree of freedom without restoring force (cf. eq. (7.9)). The procedure can be straightforwardly extended to non-abelian transformations provided that these transformations are abelian to leading order. More precisely, our solution requires that the leading order terms in the operators $\hat{G}_v$, $\hat{J}_v$ should be linear in the boson variables and should satisfy

$$[\hat{H}^{(2)}, \hat{G}_v^{(1)}] = -\frac{i}{\Im_v}\hat{J}_v^{(1)} \tag{8.19}$$

[37] This is clearly the case for the choice in eq. (2.105).

$$[\hat{G}_v^{(1)}, \hat{J}_w^{(1)}] = i\delta_{vw} \tag{8.20}$$

$$[\hat{G}_v^{(1)}, \hat{G}_w^{(1)}] = 0 \tag{8.21}$$

$$[\hat{J}_v^{(1)}, \hat{J}_w^{(1)}] = 0 \tag{8.22}$$

In a many-body problem displaying broken symmetries this can always be accomplished within the RPA, which yields both the moments of inertia $\Im_v$ and the leading order, linearised version (for instance, two quasi-particle amplitudes) of the conjugate angle $\hat{\theta}_v$ (eqs. (7.48) and (7.49)). In this approximation we may identify $\hat{H}^{(2)} = \hat{H}_{RPA}$, $\hat{G}_v^{(1)} = (\hat{\theta}_v)_{RPA}$ and $\hat{J}_v^{(1)} = (\hat{J}_v)_{RPA}$ (the RPA generators commute with each other[38]).

The existence of a solution to each of the three sets of equations requires that the system has a stable triaxial deformation. Note however that at this stage we have not yet calculated χ_{In}. Therefore the assumption of large deformations has to be verified once the variational wave function is determined. If the result of such calculations is a state that has not large triaxial deformations, then the system does not behave approximately as a triaxial rotor and the present gauge-fixing mechanism is not relevant.

The term containing the structure constants of the group in the BRST charge is also disregarded in the quadratic approximation. Therefore, for each v, the treatment of the quadratic hamiltonian and the construction of the basis corresponding to the spurious degrees of freedom proceeds as in the abelian cases.

The effects associated with the non-abelian character of the group (as well as the problems associated with the overcompletness of the basis) pertain to higher orders of perturbation theory and may be treated as such in a systematic way within the present formalism. As in the BCS case, we must choose an expansion parameter. A suitable parameter for well deformed systems is κ, the inverse of the square root of an expectation value of the quadrupole moment in the intrinsic system. The discussion in subsection 7.2.3 relative to the $O(\kappa)$ of the different terms in one- and two-body operators may be reproduced as well here. Diagrams of the same $O(\kappa)$ should be grouped together.

A possible procedure[30] is to treat perturbatively the intrinsic degrees of freedom and to perform an exact diagonalization within the collective subspace. This type of solution appears in some perturbation formalisms, in which the total Hilbert space is divided into "valence" and "complementary" subspaces. In section 9 we justify the procedure for our case of collective coordinates.

8.2 The axially symmetric case

Let us now concentrate on the case of systems having an axially symmetric mean-field approximation [39]. If our (approximate) sero-order solution preserves axial symmetry, the generator $\hat{J}_3$ does not contain linear bosonic terms

[38] In subsection 8.3 we discuss the problem of rotations at high angular momenta, in which this is not the case and thus eq. (8.19) has to be modified.

[39] However, this procedure is not followed in subsection 8.3.

and therefore the construction of the corresponding linear conjugate operator through eqs. (8.19) and (8.20) is not longer possible.

Whatever their definition, the physical meaning of the functions $\hat{G}_v$ is that the vanishing of their expectation values defines the intrinsic system. However, for an axially symmetric case it is not possible to treat collectively the rotation around the symmetry axis, because the frequency of this rotation would be the same as the one of the corresponding vibration. In the present framework this means that we should not fix the orientation of the moving frame with respect to rotations around this axis. These considerations lead us to define[40] a different gauge-fixing function $\hat{\rho}$, namely

$$\hat{\rho} = \hat{\pi}_p \hat{\Omega}_p + \hat{\eta}_p(\frac{1}{B}\hat{G}_p - \frac{A}{2B^2}\hat{P}_p) \qquad p = 1,2 \tag{8.23}$$

The resulting BRST hamiltonian

$$\begin{aligned} \hat{H}_{BRST} &= \hat{H} - \hat{\Omega}_p(\hat{J}_p - \hat{I}_p) + \frac{1}{B}\hat{G}_p\hat{P}_p - \frac{A}{2B^2}\hat{P}_p^2 \\ &+ i\hat{\bar{\pi}}_p\hat{\pi}_p + \frac{1}{B}\hat{\bar{\eta}}_v\hat{\eta}_p[\hat{G}_p, \hat{J}_v] \\ &+ i\epsilon_{pq3}\hat{\Omega}_p(\hat{\bar{\pi}}_q\hat{\eta}_3 - \hat{\bar{\pi}}_3\hat{\eta}_q) \end{aligned} \tag{8.24}$$

does not contain the gauge fixing operator $\hat{G}_3$ nor the operators $\hat{\Omega}_3$, $\hat{P}_3$, $\hat{\pi}_3$ and $\hat{\eta}_3$ which hence uncouple from the remaining degrees of freedom [41]. However, the ghost operators $\hat{\bar{\eta}}_3, \hat{\bar{\pi}}_3$ still appear and no kinetic term is associated with them. Thus the ghost space may be divided into two (degenerate) subspaces corresponding to the two different states of the 3-ghost. However, this degeneracy does not affect the perturbative calculations, since the ghost appear pairwise in (8.24) and thus intermediate states containing ghosts have the (finite) excitation energy of the p-ghost which has been simultaneously created.

We work with wave functions of the form (8.1), in which the intrinsic sector may now be also labelled by the eigenvalue K' of the operator $\hat{J}_3$. The construction of the basis for the spurious subsector proceeds as before for the two p-degrees of freedom.

The perturbation theory is based, as usual, in states that are annihilated by the BRST charge (to leading order). Hence, we consider unperturbed states[42] such that, to leading order[43],

$$\hat{f}_3\Psi = (\hat{J}_3 - \hat{I}_3)\Psi = 0 \tag{8.25}$$

---

[40] We assume $A_1 = A_2 = A$ and $B_1 = B_2 = B$.

[41] This may bring in regularization problems associated with the 3-direction which we neglect here.

[42] States with $K \neq K'$ do not strictly uncouple (since f_3 is not a constant of motion). As in the case of states with spurious phonons, these states appear as intermediate states in perturbation theory.

[43] We require that the vacuum state should not be annihilated by the operator $\hat{\bar{\eta}}_3$

The symmetry considerations in subsection 8.1.2 may be repeated here for the $p = 1, 2$ directions. In the case of the symmetry direction this result takes a stronger form. We can choose the gauge conditions $\hat{G}_1, \hat{G}_2$ such that[44]

$$[\hat{G}_p, \hat{J}_3] = i\epsilon_{qp3}\hat{G}_q \tag{8.26}$$

Thus we have

$$[\hat{H}_{BRST}, \hat{L}_3] = [\hat{Q}, \hat{L}_3] = 0 \tag{8.27}$$

and the physical wavefunctions can be classified simultaneously by their eigenvalues of $\hat{H}_{BRST}$ and $\hat{L}_3$. Using the argument at the end of subsection 4.1 we conclude that the only non-zero norm wave functions are annihilated by $\hat{L}_3$. Thus, as a consequence of axial symmetry, we only consider wavefunctions which are invariant with respect to arbitrary simultaneous rotations of their intrinsic and collective parts around the 3-axis.

8.2.1 A unitary transformation

In $\hat{H}_{BRST}$, the rotational-intrinsic coupling appears through the term $\hat{\Omega}_p\hat{I}_p$. In fact, by treating this coupling in second order of perturbation theory we obtain the (positive) rotational energy $I_p^2/2\Im$ (cf. eq. (7.16)). The corrections to the moment of inertia are given by higher order contributions including two vertices arising from the coupling term. Although this procedure is completely feasible, in the following we prefer to cast the hamiltonian into a more familiar form displaying from the outset rotational energies, Coriolis interactions, etc..

Let us define the transformation

$$\hat{T} = exp[i\hat{I}_p(\frac{1}{\Im}\hat{P}_p - \hat{G}_p)] \tag{8.28}$$

which changes the constraint $\hat{f}_p$ into $\hat{J}_p$ (to leading order) and thus eliminates (to this order) the rotational-intrinsic coupling $\hat{\Omega}_p\hat{I}_p$

$$\begin{aligned} (\hat{f}_p)_{eff} &= \hat{T}\hat{f}_p\hat{T}^\dagger \\ &= \hat{f}_p + i[\hat{I}_q(\frac{1}{\Im}\hat{P}_q - \hat{G}_q), \hat{f}_p] + \ldots \\ &= \hat{f}_p - i\hat{I}_q[\hat{G}_q^{(1)}, \hat{J}_p^{(1)}] + \ldots \\ &= \hat{J}_p + \ldots \end{aligned} \tag{8.29}$$

In the present subsection we assume that the collective angular momentum operators are of $O(1)$ in the scale discussed in subsections 7.2 and 8.1. According to the transformation to normal modes (7.10) and to the previous discussion concerning the order of magnitude of the different operators and of the moment

[44]This condition can be shown to be always satisfied by the linear (RPA) angles if the operator $\hat{J}_3$ is quadratic in the bosons.

of inertia, the exponent in the transformation (8.28) is of $O(\kappa)$. Thus the expansion associated with this transformation converges and we must include as many terms as required by the order of our perturbative calculation.

In particular, we keep[45] linear terms in $\hat{I}_v$ up to $O(\kappa^2)$ and quadratic terms up to $O(\kappa^4)$ in the transformed hamiltonian

$$\begin{aligned}
\hat{H}_{eff} &= \hat{T}\hat{H}_{BRST}\hat{T}^\dagger = \hat{H}_{BRST} \\
&-\hat{I}_p(\hat{\Omega}_p + \frac{1}{\Im}\hat{J}_p + i[\hat{H},\hat{G}_p] + \hat{\Omega}_q\hat{g}_{pq} + \frac{i}{\Im}\epsilon_{pq3}(\hat{\eta}_q\hat{\pi}_3 - \hat{\pi}_q\hat{\eta}_3)) \\
&-\frac{1}{\Im}\epsilon_{pq3}\hat{I}_3(\frac{1}{\Im}\hat{P}_q - \hat{G}_q)\hat{\Omega}_p \\
&+\frac{1}{2\Im}\hat{I}_p^2(1 - \frac{1}{2}\hat{g}_{pp} - \frac{1}{2}[\hat{G}_p,([\hat{H},\hat{G}_p] - i\hat{J}_p] - \epsilon_{pq3}^2(\frac{1}{\Im}\hat{P}_q - \hat{G}_q)^2) \\
&+\frac{3}{8\Im}\hat{I}_3^2(\frac{1}{\Im}\hat{P}_q - \hat{G}_q)^2 \qquad (8.30)
\end{aligned}$$

The first term proportional to $\hat{I}_p$ cancels the rotational-intrinsic coupling $\hat{\Omega}_p\hat{I}_p$ in $\hat{H}_{BRST}$, as expected.

The next contribution $(\frac{1}{\Im}\hat{J}_p + i[\hat{H},\hat{G}_p])$ is in principle also of $O(\kappa)$. However the leading order term in this contribution vanishes if the condition (8.19) is satisfied. This requirement yields the same moment of inertia $\Im$ as the Thouless-Valatin expression [36]. It reduces to the Inglis formula [37] if the hamiltonian $\hat{H}$ is approximated by a single-particle, deformed field[46]. All the terms proportional to $\hat{I}_v$ remaining in eq. (8.30) are of $O(\kappa^2)$. This implies a decrease of an order of magnitude with respect to the original coupling $\hat{\Omega}_p\hat{I}_p$.

The term $-\frac{1}{\Im}\hat{I}_p\hat{J}_p$ is the usual Coriolis interaction. Usually the linear (RPA) terms of this interaction are neglected on the grounds that they give rise to the moment of inertia trough the Inglis prescription and should not be counted twice (they connect, for instance, the ground state with two-quasi particle states). This neglect is justified by eq. (8.30) since those linear terms are cancelled by the corresponding terms in $[\hat{H},\hat{G}_p]$.

The Coriolis interaction is usually explicitly taken into account through the remaining (non linear) terms in the operator $\hat{J}_p$ (for instance, through the terms conserving the number of quasi-particles). A skeleton in the nuclear physics closet has been the (so far unexplained) attenuation of the Coriolis term. The hamiltonian (8.30) may solve this riddle, since it includes also non-linear terms from $i[\hat{H},\hat{G}_p]$. In fact, in the Elliot SU(3) model [38], the Casimir operator is proportional to $\hat{H} - \frac{1}{2\Im}\hat{J}_v^2$. Thus the Coriolis interaction is exactly cancelled if one uses the non-diagonal components of the quadrupole moment (which are

[45]We have assumed for simplicity that the operators $\hat{G}_p$ commute among themselves and we have omitted terms proportional to $[\hat{I}_v,\hat{I}_w]_+$ since they do not enter in the following considerations on the rotational parameters.

[46]However this last approximation is not justified within our expansion in powers of κ.

generators of the SU(3) group) as gauge-fixing operators (cf. eq. (2.105). This cancellation is to be expected, since the rotational motion decouples from the remaining degrees of freedom in Elliot's model.

On the other hand, in the field approximation for the hamiltonian $\hat{H}$, the commutation with $\hat{G}^{(1)}$ yields only linear terms, and thus the non-linear terms in the Coriolis interaction would remain unaltered. Expression (8.30) should be able to systematically treat the Coriolis effects in realistic situations.

The only contribution of $O(\kappa^2)$ to the rotational energy is given by the Thouless-Valatin moment of inertia. The corrections to this energy are of $O(\kappa^4)$ and appear from three different origins: a) the second order perturbative calculation of the terms linear in $\hat{I}_p$; b) the combination of quadratic terms $\hat{I}_p^2$ of $O(\kappa^3)$ with terms of $O(\kappa)$ from $\hat{H}_{BRST}$; and c) the expectation value of quadratic terms of $O(\kappa^4)$. Note the correspondence with the orders of magnitude and the diagrammatic corrections for the Mexican hat potential. There are no corrections to the rotational spectrum up to the $O(\kappa^6)$ (for instance, terms proportional to $\hat{I}_p^4$).

8.3 The triaxial systems at high spins

We now study systems with well-established triaxial deformations for large values of I. This can be systematically done only because we have the symmetry with respect to the (simple) collective rotation operators.

Consider the hamiltonian (4.18). Let us start by interpreting the $\hat{\Omega}_v$ as the components of the angular velocity operator by calculating

$$[\hat{H}_{BRST}, \hat{I}_v] = i\, \epsilon_{vws} \hat{\Omega}_w \hat{I}_s \tag{8.31}$$

that is

$$[\hat{H}_{BRST}, \vec{I}] = i(\vec{\Omega} \wedge \vec{I}) \tag{8.32}$$

This equation is also valid for a rigid rotor (cf. appendix 8.A), where

$$\hat{\Omega}_v \equiv \hat{I}_v / \Im_v \tag{8.33}$$

In our general case a relation like (8.33) only appears a posteriori. At this stage *the* $\hat{I}_v$ *and the* $\hat{\Omega}_v$ *are independent operators.*

We may factorize the wave functions as follows:

$$\Psi_{IM} = |coll>_{IM}\ |intr>\ |Lagr.\ mult>\ |ghosts> \tag{8.34}$$

Now, from the discussion of discrete symmetries, the trial wave functions should be of the form

$$\Psi_{IM}^{(S)} = \frac{1}{4}(1 + \hat{\mathcal{R}}_1 + \hat{\mathcal{R}}_2 + \hat{\mathcal{R}}_3)\Psi_{IM} \tag{8.35}$$

so that

$$\hat{R}_v \Psi_{IM}^{(S)} = \Psi_{IM}^{(S)} \tag{8.36}$$

However, if I *is large and the deformations of the resulting system turns out to be large* (i.e. we are in a sort of semiclassical limit) we can approximately break the symmetry (8.36) and work with wavefunctions of the form (8.34). This is in direct analogy with what is done in the treatment of the rigid rotor in the Holstein-Primakoff representation that is discussed in appendix 8.A.

Notice that

$$< \Psi_{IM}^{(S)} | \hat{I}_v | \Psi_{IM}^{(S)} >= 0 \tag{8.37}$$

while

$$< \Psi_{IM} | \hat{I}_v | \Psi_{IM} > \neq 0 \tag{8.38}$$

for some v. Again, the breaking of the symmetry (8.36) is essential for having non-zero expectation values of angular momentum operators (see the discussion in appendix 8.A).

The minimization equations (4.19) imply in this case

$$< \hat{P}_v >=< \hat{G}_v >=< \vec{\Omega} \wedge \vec{I} >= 0 \tag{8.39}$$

Since we attempt to work with large values of I we use the Holstein-Primakoff representation for the collective angular momenta. At this level we only keep the leading term in powers of I^{-1} (or alternatively $(I + \frac{1}{2})^{-1}$, see footnote below) so that the last equation becomes (cf. appendix 8.A)

$$< \vec{\Omega} \wedge \vec{I} >= I < \vec{\Omega} \wedge \vec{v}_x >= 0$$

where the directions $(\vec{v}_x, \vec{v}_y, \vec{v}_z)$ are yet unspecified and $\vec{I} \simeq I\vec{v}_x$.

Finally, we obtain

$$< \hat{\Omega}_y >=< \hat{\Omega}_z >=< \hat{I}_y >=< \hat{I}_z >=< \hat{J}_y >=< \hat{J}_z >= 0 \tag{8.40}$$

$$< \hat{\Omega}_x >= \Omega \quad < \hat{J}_x >=< \hat{I}_x >= I \tag{8.41}$$

where Ω is a *constant*: the expectation value of the norm of the angular velocity vector[47]. We define the moment of inertia $\Im$ as the ratio

$$\Im \equiv \frac{I}{\Omega} \tag{8.42}$$

Now, denoting by δ_{intr} a variation with respect to the intrinsic wavefunctions we have, using the product form for the wavefunctions.

[47] Eq. (8.41) implies the choice of I^{-1} as the semiclassical high spin expansion parameter although there are indications that the choice $(I + 1/2)^{-1}$ is more natural. In that case the last of (8.41) would read

$$< \hat{J}_x >=< \hat{I}_x >= I + 1/2$$

However, in what follows we use I for simplicity.

$$\delta_{intr} < |\hat{H}_{BRST}| >= \delta_{intr}[< intr|H|intr > -\Omega < intr|\hat{J}_x|intr >] = 0 \quad (8.43)$$

A cranked minimization has naturally emerged for the intrinsic part of the wavefunctions. The usual c-number Ω has appeared as an expectation value of an angular velocity operator.

We are now faced with the task of determining the cranking direction $\vec{v}_x$. Details of the different cranking possibilities are given in [16]. Here we consider the case such that the directions v_x and 1 coincide with each other. Eq. (8.43) in this case reads:

$$\delta < intr|H - \Omega\hat{J}_1|intr >= 0 \quad (8.44)$$

It is easy to see that $H - \Omega\hat{J}_1$ is invariant under (intrinsic) 180-degree rotations around the 1-axis. In what follows, we assume that $|intr>$ also has this discrete symmetry (i.e. that the minimization does not break this symmetry). This fact, in turn, implies that the variational vacuum in the product space is symmetric with respect to the operator $\hat{\mathcal{R}}_1$ (cf. (8.8)) in this space. We will use this result in the next section.

Therefore the cranking approximation has emerged from the minimization procedure, under the assumptions that the system has a mean field with well established triaxial deformations and that we are working with reasonably high values of I. (i.e. values of I such that the expansion in appendix 8.A is good, see condition quoted there). We henceforth assume that the minimization (8.44) yields the self-consistent cranked Hartree wave function[48].

8.3.1 The quadratic hamiltonian

In order to obtain the normal modes associated with the BRST hamiltonian we must generalize the usual RPA procedure. In the first place we obtain a set of independent bosons. Secondly, we diagonalize the quadratic expression of the BRST hamiltonian in terms of these bosons and the associated fermion operators.

A part of this program has already been carried out in refs. [40,41] which discuss the self-consistent cranked Hartree-Bogoliubov approximation plus the RPA treatment of the routhian $H - \Omega\hat{J}_1$. We use these results whenever relevant for our formulation.

The magnitudes entering in the previous minimization of the hamiltonian are to be expanded about their value at equilibrium. This expansion is trivial for the boson modes $(\hat{\Omega}_v, \hat{P}_v)$ (cf. eqs. (8.40) and (8.41)).

[48]The choice of Hartree instead of Hartree-Fock minimization does not follow necessarily from our derivation. It is, however, the most natural choice for treatments applying bosonization procedures or the nuclear field theory.

$$\hat{\Omega}_v = \delta_{\nu 1}\Omega + \hat{\Omega}_v^{(1)}$$
$$\hat{P}_v = \hat{P}_v^{(1)} \tag{8.45}$$

where $\hat{\Omega}_v^{(1)}$ and $\hat{P}_v^{(1)}$ are conjugate boson operators with vanishing expectation values.

In the case of the microscopic angular operators $\hat{J}_v$ we write

$$\hat{J}_v = \delta_{v1} I + \hat{J}_v^{(1)} + \hat{J}_v^{(2)} + \hat{J}'_v \tag{8.46}$$

where $\hat{J}_v^{(1)}$ and $\hat{J}_v^{(2)}$ are linear and quadratic expressions in the boson operators respectively, and $\hat{J}'_v$ contains the remaining terms (if any). In the basis space defined by the routhian (8.44), the following relations are satisfied

$$[\hat{J}_v^{(1)}, \hat{J}_w^{(1)}] = i\epsilon_{vw1} I \tag{8.47}$$

The construction of the normal modes only requires the linear term of the gauge fixing function $\hat{G}_v$. We now choose them to be equal to the intrinsic RPA angles[49] $\hat{G}_v = (\hat{\theta}_v)_{RPA}$ satisfying the relations (8.20)-(8.22) and

$$[(\hat{\theta}_v)_{RPA}, \Gamma_r^\dagger] = 0 \tag{8.48}$$

where $\Gamma_r^\dagger$ are the creation operators for all the other (finite frequency) normal modes[50]. Note that the $\hat{G}_v$ defined as the linear part of the gauge conditions that are choosen as in eq. (2.105) satisfy (8.20) and (8.21) but do not commute with all the other normal modes. This is the practical disadvantage of that choice, which otherwise is also correct.

Finally, the components of the collective angular momentum may be bosonized, for instance, with the Holstein-Primakoff representation (cf. appendix 8.A)

$$\hat{I}_v = I\delta_{v1} + \hat{I}_v^{(1)} + \hat{I}_v^{(2)} + \hat{I}'_v \tag{8.49}$$

such that

$$\hat{I}_1^{(1)} = \hat{I}_2^{(2)} = \hat{I}_3^{(2)} = 0 \tag{8.50}$$

and

$$[\hat{I}_v^{(1)}, \hat{I}_w^{(1)}] = -i\epsilon_{\nu w1} I \tag{8.51}$$

The pairs of operators

$$(\theta_v^{(1)}, J_v^{(1)}), (\hat{\Omega}_v^{(1)}, \hat{P}_v^{(1)}), (\hat{I}_3^{(1)}/\sqrt{I}, \hat{I}_2^{(1)}/\sqrt{I}) \tag{8.52}$$

---

[49]The usual RPA equation determining both the angle and the moment of inertia can be applied here to the rotation around the $v = 1$ axis. The generalization applicable to the case of rotations around the $v = 2, 3$ axis is given in eq. (8.63) (cf. ref. [16]).

[50]We assume that the part of the hamiltonian concerning these other degrees of freedom has been already diagonalized (within the RPA), and we omit them from the quadratic hamiltonian in the remaining part of this section.

are conjugate to each other but they are not mutually independent, since for instance $[\hat{J}_2^{(1)}, \hat{J}_3^{(1)}]$ does not vanish (cf. eq.(8.47)).

A convenient set of independent operators is constituted by the conjugate pairs

$$(\hat{\Omega}_v^{(1)}, \hat{P}_v^{(1)}), (\hat{\theta}_1^{(1)}, \hat{J}_1^{(1)}), (\hat{\theta}_2^{(1)}, \hat{f}_2^{(1)}), (\hat{\theta}_3^{(1)}, \hat{f}_3^{(1)}), (\hat{s}, \hat{t}) \tag{8.53}$$

where $\hat{f}_v^{(1)}$ is the linear part of the constraint and

$$\begin{aligned} \hat{s} &\equiv \hat{I}_3^{(1)}/\sqrt{I} + \hat{\theta}_2^{(1)}\sqrt{I} \\ \hat{t} &\equiv \hat{I}_2^{(1)}/\sqrt{I} - \hat{\theta}_3^{(1)}\sqrt{I} \end{aligned} \tag{8.54}$$

The operators (8.53) satisfy the non vanishing commutation relations

$$[\hat{s}, \hat{t}] = [\hat{\theta}_1^{(1)}, \hat{J}_1^{(1)}] = [\hat{\theta}_2^{(1)}, \hat{f}_2^{(1)}] = [\hat{\theta}_3^{(1)}, \hat{f}_3^{(1)}] = [\hat{\Omega}_v^{(1)}, \hat{P}_v^{(1)}] = i, \tag{8.55}$$

all other commutators being zero.

The quadratic part of the hamiltonian can be split in two terms if use is made of the symmetry with respect to the rotation operator $\hat{\mathcal{R}}_1$. This is an extension of what is done in [40,41] for the intrinsic part. The two terms correspond to rotations around the $v = 1$ axis and rotations around the $v = 2, 3$ axis. They are labelled with the subindices x and $\perp$, respectively.

Rotations around the $v = 1$ axis

The corresponding terms in the quadratic hamiltonian (8.56) are written

$$\hat{H}_x^{(2)} = \hat{H}_{ax}^{(2)} + \hat{H}_{bx}^{(2)} + \hat{H}_{gx}^{(2)} \tag{8.56}$$

where

$$\begin{aligned} \hat{H}_{ax}^{(2)} &= (\hat{J}_1^{(1)})^2/2\Im_1 \\ \hat{H}_{bx}^{(2)} &= -\hat{\Omega}_1^{(1)}\hat{J}_1^{(1)} + \frac{1}{B_1}\hat{\theta}_1^{(1)}\hat{P}_1^{(1)} - \frac{A_1}{2B_1^2}(\hat{P}_1^{(1)})^2 \\ \hat{H}_{gx}^{(2)} &= i\hat{\pi}_1\hat{\bar{\pi}}_1 + \frac{i}{B_1}\hat{\eta}_1\hat{\bar{\eta}}_1 \end{aligned} \tag{8.57}$$

The moment of inertia $\Im_1$ and the linear angle $\hat{\theta}_1^{(1)}$ are obtained from eqs. (8.19) and (8.20). As mentioned before, these equations yield the Thouless-Valatin moment of inertia $\Im_1$ which is in general different from $\Im$ (8.42). The solution for the spurious sector presented in section 7 can be straightforwardly applied.

Rotations around the $v = 2, 3$ axes

These degrees of freedom are coupled in the remaining terms of the hamiltonian

$$\hat{H}_\perp^{(2)} = \hat{H}_{a\perp}^{(2)} + \hat{H}_{b\perp}^{(2)} + \hat{H}_{g\perp}^{(2)} \tag{8.58}$$

$$
\begin{aligned}
\hat{H}^{(2)}_{a\perp} &= \hat{H}^{(2)}_{\perp} - \Omega \hat{f}^{(2)}_1 \\
\hat{H}^{(2)}_{b\perp} &= -\hat{\Omega}^{(1)}_2 \hat{f}^{(1)}_2 - \hat{\Omega}^{(1)}_3 f^{(1)}_3 + \frac{1}{B_2}\hat{\theta}^{(1)}_2 \hat{P}^{(1)}_2 + \frac{1}{B_3}\hat{\theta}^{(1)}_3 \hat{P}^{(1)}_3 \\
&\quad - \frac{A_2}{2B_2^2}(\hat{P}^{(1)}_2)^2 - \frac{A_3}{2B_3^2}(\hat{P}^{(1)}_3)^2 \\
\hat{H}^{(2)}_{g\perp} &= i\hat{\pi}_2\hat{\bar{\pi}}_2 + i\hat{\pi}_3\hat{\bar{\pi}}_3 + \frac{i}{B_2}\hat{\eta}_2\hat{\bar{\eta}}_2 + \frac{i}{B_3}\hat{\eta}_3\hat{\bar{\eta}}_3 \\
&\quad + i\Omega(\hat{\bar{\pi}}_2\hat{\eta}_3 - \hat{\bar{\pi}}_3\hat{\eta}_2)
\end{aligned} \tag{8.59}
$$

Before attempting to find the normal modes we must generalize eqs.(8.19), which would apply for the $v = 2, 3$ axis only in the limit $\Omega \to 0$. We obtain the following relations by conserving only the linear terms in both sides of the equations (as in (8.19))

$$[\hat{H}^{(2)}_{a\perp}, \hat{f}^{(1)}_2] = -i\Omega \hat{f}^{(1)}_3$$

$$[\hat{H}^{(2)}_{a\perp}, \hat{f}^{(1)}_3] = i\Omega \hat{f}^{(1)}_2$$

$$[\hat{H}^{(2)}_{a\perp}, (\hat{s} - \hat{\theta}^{(1)}_2\sqrt{I})] = i\Omega(\hat{t} + \hat{\theta}^{(1)}_3\sqrt{I})$$

$$[\hat{H}^{(2)}_{a\perp}, (\hat{t} + \hat{\theta}^{(1)}_3\sqrt{I})] = i\Omega(-\hat{s} + \hat{\theta}^{(1)}_2\sqrt{I}) \tag{8.60}$$

$\hat{H}_\perp$ commutes both with the constraints and with the components of the collective angular momentum.

The most general quadratic hamiltonian satisfying the commutation relations (8.60) plus the condition that the coefficient of the term $\hat{f}_2\hat{f}_3$ should vanish (condition of principal axis) is

$$
\begin{aligned}
\hat{H}^{(2)}_{a\perp} &= \frac{1}{\Im_2}[\frac{1}{2}(\hat{f}^{(1)}_2)^2 + \sqrt{I}\hat{t}\hat{f}^{(1)}_2] + \frac{1}{\Im_3}[\frac{1}{2}(\hat{f}^{(1)}_3)^2 + \sqrt{I}\hat{s}\hat{f}^{(1)}_3] \\
&\quad + \Delta_3^2 I\hat{s}^2 + \Delta_2^2 I\hat{t}^2 + \Omega(\hat{f}^{(1)}_2\hat{\theta}^{(1)}_3 - \hat{f}^{(1)}_3\hat{\theta}^{(1)}_2)
\end{aligned} \tag{8.61}
$$

where

$$\Delta_v = (\frac{1}{\Im_v} - \frac{1}{\Im})^{1/2} \tag{8.62}$$

The hamiltonian (8.61) is a generalization of the first eq. (8.57). From it one obtains the relations

$$[\hat{H}^{(2)}_{a\perp}, \hat{\theta}^{(1)}_2] = -\frac{i}{\Im_2}(\hat{f}_2 + \sqrt{I}\hat{t}) - i\Omega\hat{\theta}^{(1)}_3 = -\frac{i}{\Im_2}\hat{J}^{(1)}_2 + iI\Delta_2^2\hat{\theta}^{(1)}_3$$

$$[\hat{H}^{(2)}_{a\perp}, \hat{\theta}^{(1)}_3] = -\frac{i}{\Im_3}(\hat{f}_3 + \sqrt{I}\hat{s}) + i\Omega\hat{\theta}^{(1)}_2 = -\frac{i}{\Im_3}\hat{J}^{(1)}_3 - iI\Delta_3^2\hat{\theta}^{(1)}_2 \tag{8.63}$$

which are a generalization of (8.19). The combination of (8.63) with the commutation relations (8.20) and (8.21) determines the moments of inertia $\Im_2, \Im_3$ and the angles $\hat{\theta}_2^{(1)}, \hat{\theta}_3^{(1)}$.

In ref. [16] the normal modes associated with the quadratic hamiltonians obtained above are explicitly given. Here we only quote the final results in terms of the normal mode operators at the linearized RPA level.

In addition to the real modes $\Gamma_r^\dagger$, the normal modes turn out to be: i) a mode $\Gamma_b^\dagger$ (which we identify with a wobbling mode) and ii) the spurious phonon modes $\Gamma_{1v}^\dagger, \Gamma_{ov}^\dagger$ plus ghost modes $\bar{a}_v, \bar{b}_v$. The corresponding phonon plus ghost hamiltonian is

$$\hat{H}_{BRST}^{(2)} = \omega_b \Gamma_b^\dagger \Gamma_b + \omega_r \Gamma_r^\dagger \Gamma_r + \tag{8.64}$$

$$+ \omega_v (\Gamma_{1v}^\dagger \Gamma_{1v} - \Gamma_{ov}^\dagger \Gamma_{ov} + \bar{a}_v a_v + \bar{b}_v b_v) \tag{8.65}$$

which displays the BRST supersymmetry within the spurious sector.

The frequency

$$\omega_b \equiv I \Delta_2 \Delta_3 \tag{8.66}$$

is the expected excitation (wobbling) energy for a triaxial rotor with moments of inertia $\Im, \Im_2, \Im_3$ (cf. appendix 8.A).

The ground state energy is given by

$$W_{BRST}^{(2)} = < intr|\hat{H}|intr > + \frac{\Omega}{2} + \frac{\omega_b}{2} + \sum_r \frac{\omega_r}{2} \tag{8.67}$$

The two first terms have been discussed in [40] and shown to yield the $\frac{1}{2\Im}I(I+1)$ rotational energy. There is also a zero-point energy associated with the physical modes (including the wobbling mode). As in previous cases, the spurious degrees of freedom do not yield any RPA contribution to the ground state energy, since the ghost and phonon constants mutually cancel.

In terms of the normal modes, the quadratic part of the BRST charge reads

$$\hat{Q}^{(2)} = \sum_v \hat{Q}^v \tag{8.68}$$

$$\hat{Q}_v^{(2)} \propto (\Gamma_{1v}^\dagger + \Gamma_{ov}^\dagger) a_v - (\Gamma_{1v} + \Gamma_{ov}) \bar{b}_v \tag{8.69}$$

We see that the ground state defined by the absence of phonons and ghosts is annihilated by the operator $\hat{Q}^{(2)}$. Therefore it is a physical state (to this order). It carries good values of the quantum numbers I, M and has the maximum projection I along the $v = 1$ axis (minus the harmonic correlation). It is also straightforward to check that

$$[\hat{Q}^{(2)}, \Gamma_r^\dagger] = [\hat{Q}^{(2)}, \Gamma_b^\dagger] = 0 \tag{8.70}$$

while

$$[\hat{Q}^{(2)}, \Gamma^{\dagger}_{0v}] \neq 0 \quad [\hat{Q}^{(2)}, \Gamma^{\dagger}_{1v}] \neq 0 \tag{8.71}$$

$$[\hat{Q}^{(2)}, \bar{a}_v] \neq 0 \quad [\hat{Q}^{(2)}, \bar{b}_v] \neq 0 \tag{8.72}$$

Eqs. (8.70) imply that states created by the operators $\Gamma^{\dagger}_r$ and $\Gamma^{\dagger}_b$ are physical (as they should), while eqs. (8.71) and (8.72) show that states having $(1v)$, $(0v)$ phonons or ghosts are unphysical. Thus the word *spurious* has a definite meaning here.

A perturbative calculation with the BRST residual hamiltonian $\hat{H}_{res}$

$$\hat{H}_{res} = \hat{H}_{BRST} - \hat{H}^{(2)}_{BRST} \tag{8.73}$$

is free from the problems associated with broken symmetries. The perturbative calculation requires the replacement of the boson operators in $\hat{H}_{res}$ by their normal-mode counterparts $\Gamma^{\dagger}_{nv}, \Gamma_{nv}, \Gamma^{\dagger}_{b}, \Gamma_{b}, \Gamma^{\dagger}_{r}, \Gamma_{r}$. The construction of the vertices is made as in the usual systematic procedures to correct the RPA.

We have not supposed that the particle number is violated in the minimization procedure. However, this additional breakdown of a symmetry would not add a serious complication. We should combine the procedures presented in this subsection with those in subsection 7.2.

The perturbative corrections to the RPA are practically not more complicated than for the original hamiltonian, since the inclusion of three extra phonons with a negative metric and of six ghosts is simply accomplished from the computational point of view.

appendix 8.A The high spin treatment of the asymmetric rotor

We use the high-spin treatment of the asymmetric rotor given in [2] in order to illustrate the partial violation of the 180-degree rotation symmetries (subsection 8.3) and in order to introduce the boson mapping for the collective components of the angular momentum which satisfy (8.51).

The quantum rotor hamiltonian is

$$\hat{H}_{rot} = \frac{\hat{I}_v^2}{2\Im_v} \quad (\Im_x > \Im_y > \Im_z) \tag{8.74}$$

$\hat{H}_{rot}$ commutes with I^2 and with $\hat{I}'_z$. The eigenfunctions are of the form

$$\Psi_{IM} = \sum_{K=-I}^{K=+I} c_K D^I_{MK} \tag{8.75}$$

where K are the eigenvalues of $\hat{I}_x$.

We assume for simplicity that I is an integer. $\hat{H}_{rot}$ is also invariant with respect to the discrete group of transformations given by the rotation operators $\hat{R}^I_v$. We can further classify states according to the representations of this group [2]. The states are split into four families:

$$\begin{aligned}
\Psi_{IM\pm} &= \sum_{K=-I}^{K=+I} c_K (D^I_{MK} \pm D^I_{M-K}) \quad K \text{ odd} \\
\Psi_{IM\pm} &= \sum_{K=-I}^{K=+I} c_K (D^I_{MK} \pm D^I_{M-K}) \quad K \text{ even}
\end{aligned} \tag{8.76}$$

We now want to approximate the eigenfunctions of (8.75) within a representation I, M for large I. We use the Holstein-Primakoff representation

$$\begin{aligned}
\hat{I}_+ &= \hat{I}_y + i\hat{I}_z = \hat{I}_-^\dagger = \alpha_3^\dagger\sqrt{2I - \alpha_3^\dagger\alpha_3} \simeq \sqrt{2I}\alpha_3^\dagger \\
\hat{I}_x &= I - \alpha_3^\dagger\alpha_3 \\
\left[\alpha_3, \alpha_3^\dagger\right] &= 1
\end{aligned} \tag{8.77}$$

$\hat{H}_{rot}$ can now be written including terms up to order I^0 and a quadratic hamiltonian in the bosons is obtained. In normal modes it has the form

$$\hat{H}_{rot} = \frac{I(I+1)}{2\Im_x} + \omega_b(n + \frac{1}{2}) \tag{8.78}$$

where

$$\omega_b = I\left[\left(\frac{1}{\Im_z} - \frac{1}{\Im_x}\right)\left(\frac{1}{\Im_y} - \frac{1}{\Im_x}\right)\right]^{1/2} \tag{8.79}$$

According to ref. [2] this approximation holds if

$$I >> \left[\left(\frac{1}{\Im_z} - \frac{1}{\Im_x}\right) + \left(\frac{1}{\Im_y} - \frac{1}{\Im_x}\right)\right]\left[\left(\frac{1}{\Im_z} - \frac{1}{\Im_x}\right)\left(\frac{1}{\Im_y} - \frac{1}{\Im_x}\right)\right]^{-1/2} \tag{8.80}$$

Each quantum of excitation decreases by one unit the projection of the angular momentum along the x-axis. Therefore the separation between even and odd number of bosons corresponds to the separation between even and odd values of K. Both are a consequence of the symmetry under the transformation $\mathcal{R}_x^I$ which is unbroken. However, we no longer work with combinations of states with $\pm K$, due to the fact that the symmetry associated with $\mathcal{R}_y^I, \mathcal{R}_z^I$ is broken. To leading order we have that the expectation value in the ground state:

$$< 0|\hat{I}_x|0 > \simeq I \tag{8.81}$$

while in the representation (8.77)

$$< IM|\hat{I}_x|IM > = 0 \tag{8.82}$$

because of the symmetry under $\mathcal{R}_y^I$ and $\mathcal{R}_z^I$.

This is in analogy with the semiclassical limit for a particle in a symmetric double well: the ground state wavefunction peaks sharply around the two minima and the expectation value of the operator x is zero. However, approximations based on small oscillations around one of the minima yield good results in the semiclassical regime; for them the expectation value of x is non zero. The symmetry is of course restored by tunneling between solutions peaked on either minimum, but this effect is exponentially small.

9 The collective dynamics

In the previous sections we have constructed Hamiltonians that operate on a space that includes collective coordinates (ϕ_v) as well as intrinsic coordinates.

A possible course to follow (although not the only one, cf. subsection 8.3) is to "eliminate" the intrinsic subspace and thus obtain a collective hamiltonian operating purely within the collective subspace. This elimination may be carried out with a standard technique, which in the context of functional integrals amounts to integrate over the intrinsic variables. In practice, this can only be done perturbatively by adding a number of "intrinsic diagrams" up to a certain level of accuracy. The parameters of the collective hamiltonian operator thus obtained are dependent on the intrinsic structure.

The case of nuclear rotations is paradigmatic: we start with a microscopic hamiltonian and we introduce collective coordinates defining the orientation of a rotating frame that is fixed to the system by the gauge conditions. From the effective (BRST) hamiltonian that operates in the product space one calculates a collective hamiltonian operator describing a quantum rotor (plus higher powers in the collective angular momentum) whose parameters of inertia are dependent on the intrinsic structure.

In this section we give a derivation that is appropriate for the collective dynamics associated with the intrinsic ground state. The generalisation introduces no serious problems.

Let us consider a basis of the Hilbert space the product form

$$|a_i > \otimes |b_r > \equiv |\Psi_{ir} > \tag{9.1}$$

where the $|b_r >$ are functions of the collective variables ϕ_v and $|a_i >$ are functions of the intrinsic variables.

The effective evolution operator is

$$U(\beta) = exp[-\beta \hat{H}] \tag{9.2}$$

where $\hat{H}$ is the effective hamiltonian, $\beta \equiv \tau_f - \tau_i$ and $\tau \equiv it$. For any operator $\hat{A}$

$$< \hat{A} >^{(\beta)} = \frac{tr\ U\hat{A}}{tr\ U} = \frac{\sum exp[-\beta\epsilon_l] < \Psi_l|\hat{A}|\Psi_l >}{\sum exp[-\beta\epsilon_l] < \Psi_l|\Psi_l >} \tag{9.3}$$

where ϵ_l, $|\Psi_l >$ are the exact energies and eigenstates.

In particular, if the vacuum $|\Psi_o >$ is non degenerate:

$$\lim_{\beta\to\infty} < \hat{A} >^{(\beta)} = \frac{< \Psi_o|\hat{A}|\Psi_o >}{< \Psi_o|\Psi_o >} \tag{9.4}$$

and

$$\lim_{\beta\to\infty} \frac{1}{\beta}\ log\ tr\ U = \epsilon_o \tag{9.5}$$

Now, using the basis (9.1) we have that:

$$tr\ U = \sum_{i,r} < b_r| < a_i|U|a_i > |b_r > \tag{9.6}$$

and

$$< A >^{(\beta)} = \frac{\sum_{i,r} < b_r| < a_i|U\hat{A}|a_i > b_r >}{tr\ U} \tag{9.7}$$

Let us now define

$$\begin{aligned} U^{coll} &= \sum_i < a_i|U|a_i > \\ < b_r|U^{coll}|b_s > &\equiv < b_r|\sum_i < a_i|U|a_i > |b_s > \end{aligned} \tag{9.8}$$

where we have taken a "partial" trace. Clearly,

$$tr\ U(\beta) = tr^{coll}\ U^{coll} = \sum_r < b_r|U^{coll}|b_r > \tag{9.9}$$

where tr^{coll} is a trace over collective states, since (9.9) is just another way of writing (9.7).

If $\hat{B}$ is an operator that acts only on the collective part of the wavefunctions

$$\begin{aligned} < \hat{B} >^{(\beta)} &= \frac{\sum_{i,r} < b_r| < a_i|U(\beta)|a_i > \hat{B}|b_r >}{tr^{coll}\ U^{coll}} \\ &= \frac{\sum_r < b_r|U^{coll}\hat{B}|b_r >}{tr^{coll}\ U^{coll}} = \frac{tr^{coll}\ (U^{coll}\hat{B})}{tr^{coll}\ U^{coll}} \end{aligned} \tag{9.10}$$

The equations (9.9) and (9.10) tell us that we can interpret U^{coll} as a collective evolution operator, and that

$$\tilde{H}^{coll} \equiv \frac{1}{\beta}\ log\ U^{coll} \longleftrightarrow U^{coll} = exp\ [-\beta\tilde{H}^{coll}] \tag{9.11}$$

defines $\tilde{H}^{coll}$ as a collective hamiltonian.

Clearly, if we need to calculate the expectation value of an operator that acts on intrinsic as well as collective wavefunctions, we cannot use the r.h.s. of (9.10) and have to revert to (9.7).

We also define for future use H^{coll} :

$$\begin{aligned} H^{coll} &= lim_{\beta\to\infty}\ \tilde{H}^{coll} = lim_{\beta\to\infty}\ -\frac{1}{\beta}\ log\ U^{coll} \\ < b_l|H^{coll}|b_m > &= lim_{\beta\to\infty}[\ log\ (< b_s|\sum < a_i|U|a_i > |b_r >)\ ]_{lm} \end{aligned} \tag{9.12}$$

It is this expression that we calculate perturbatively by adding diagrams involving the intrinsic part.

The matrix elements of the evolution operator can be written

$$< b_l|U^{coll}(\beta)|b_m >= \sum_{v,w} < b_l|\phi_v >< \phi_v| < a_i|U|a_i > |\phi_w >< \phi_w|b_m > \tag{9.13}$$

where we have inserted two collective identities $\sum_v |\phi_v >< \phi_v|$. Let us now represent a part of this equation by means of a functional integral

$$Z =< \phi_r|\sum_i < a_i|U|a_i > |\phi_s >= \int D[\phi_v, P_v]\, D[z_i, z_i^*]\, exp[S_{eff}] \tag{9.14}$$

The integral over intrinsic fermion and/or boson variables z_i is the familiar functional expression for the trace of the evolution operator (cf. ref. [31]). Antiperiodical (periodical) limits for fermion (boson) variables are assumed for the limits of the intrinsic part integrals while for the collective part the limits are ϕ_r, ϕ_s. The effective action S_{eff} can be written

$$S_{eff} = \int (\mathcal{L}_o + \sum_v \dot{\phi}_v P_v - H'')\, dz \tag{9.15}$$

where $\mathcal{L}_o$ includes the sero order intrinsic part (free fermions and/or bosons)

$$\mathcal{L}_o = \sum_i (z_i^* \dot{z}_i - \omega_i z_i^* z_i) \tag{9.16}$$

and all the rest is included in H''. *We are not including any collective variables in $\mathcal{L}_o$, since we do not want to treat them perturbatively.*

We now proceed to construct the perturbative series in the usual way. Let us write (9.14) in the following way

$$\begin{aligned} Z = \quad & \int D[\phi_v, P_v] D[z_i, z_i^*]\, exp\, [\int (\mathcal{L}_o + \textstyle\sum_w \dot{\phi}_w P_w) d\tau] \\ & \times (1 + \int H''(\tau_1) d\tau_1 + \tfrac{1}{2!} \int H''(\tau_1) H''(\tau_2) d\tau_1 d\tau_2 + ...) \end{aligned} \tag{9.17}$$

Consider, for example, a second order term of the form

$$\begin{aligned} & \int D[\phi_v, P_v] D[z_i, z_i^*]\, exp[\int (\mathcal{L}_o + \textstyle\sum_w \dot{\phi}_w P_w) d\tau] \\ & \times \int d\tau_1 d\tau_2\, B_1^{coll}(\tau_1) z_j^*(\tau_1) B_2^{coll}(\tau_2) z_j(\tau_2) \end{aligned} \tag{9.18}$$

where B_i^{coll} are purely collective operators. This term is

$$\begin{aligned} & \int d\tau_1 d\tau_2\, \{\int D[\phi_v, P_v]\, B_1^{coll}(\tau_1) B_2^{coll}(\tau_2)\, exp[-\int \textstyle\sum_w \dot{\phi}_w P_w d\tau] \\ & \times \int D[z_i, z_i^*]\, z_j^*(\tau_1) z_j(\tau_2)\, exp[-\int \mathcal{L}_o d\tau]\} \end{aligned} \tag{9.19}$$

The second factor in (9.19) is simply Feynman's diagram $< z_j^*(\tau_1), z_j(\tau_2) >$ (cf. diagram (a) in fig. 5), while the first one is

$$\begin{aligned} & < \phi_r|\hat{B}_1^{col} \hat{B}_2^{col}|\phi_s > \qquad \text{if} \quad \tau_1 > \tau_2 \\ & < \phi_r|\hat{B}_2^{col} \hat{B}_1^{col}|\phi_s > \qquad \text{if} \quad \tau_2 > \tau_1 \end{aligned} \tag{9.20}$$

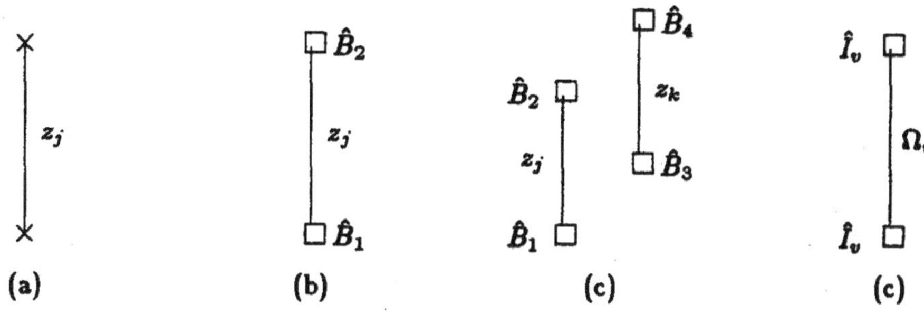

Figure 5: Some Feynman's diagrams appearing in the perturbation theory associated with collective dynamics.

This can be understood by observing that the unperturbed collective part of the evolution operator is simply the identity operator (there are no collective terms apart from $\sum_w \dot{\phi}_w P_w$ in the unperturbed action of (9.18), (9.19)). Putting both things together we have this particular element of (9.13) that corresponds to this second order term

$$\int \quad [< b_l|\hat{B}_1^{coll}\hat{B}_2^{coll}|b_m > \theta(\tau_1 - \tau_2) + < b_l|\hat{B}_2^{coll}\hat{B}_1^{coll}|b_m > \theta(\tau_2 - \tau_1)] \\ \times < z_1^*(\tau_j)z_j(\tau_2) > \; d\tau_1 d\tau_2 \tag{9.21}$$

(where θ is the step function) which we represent diagrammatically as in diagram (b), where the square vertices indicate that these vertices also involve a collective operator (crosses are used for vertices that do not involve collective operators).

In this construction, the functional integral over collective states only serves the purpose of time-ordering the collective operators.

Repeating this procedure for all the terms in (9.17) we obtain the matrix elements of the collective evolution operator by adding Feynman diagrams.

In order to calculate the logarithm of this operator, we would like to apply the linked cluster theorem of perturbation theory [31]. However, this theorem cannot be straightforwardly applied since the time-ordering acts as a link between collective operators, i.e. apparently unlinked diagrams with "square vertices" are not truly unlinked. For example, consider the fourth order diagram (c) that results from the square of a term like (9.19). We find that the two components do not factorize because of the time-ordered product

$$T[\hat{B}_1(\tau_1)\hat{B}_2(\tau_2)\hat{B}_1(\tau_3)\hat{B}_2(\tau_4)] \tag{9.22}$$

which itself does not factorize if the $\hat{B}_i$ do not commute.

Let us now see the more realistic example of rotations in the BRST treatment of section 8. Let us write

$$\hat{H}_{BRST} = \hat{H}_1 + \hat{\Omega}_v \hat{I}_v \tag{9.23}$$

where we have explicitly separated the only part of H_{BRST} containing collective operators.

Using the results of this section we conclude that U^{coll} is obtained (apart from constant terms) by summing diagrams which have as many Ω_v insertions as I_v operators in a given term. For example the quantum rotor term is obtained to leading order by calculating the contraction $< \hat{\Omega}_v(\tau), \hat{\Omega}_v(\tau') >$ represented by diagram (d), which is shown in section 8 to yield the Thouless-Valatin moment of inertia.

10 Conclusions

The introduction of k collective parameters which determine the basis of the quantum Hilbert space has been a common practice in many-body physics.

Instead of specifying a priori the time-dependance of such parameters we can consider them as additional variables. The wavefunctions in the resulting enlarged space will hence depend on the additional k variables, as well as on those of the original problem. In this case the resulting system posesses a gauge invariance generated by the k-constraints f_v (or F_v) and the evolution of such collective coordinates is only determined through the gauge-fixing procedure plus the dynamics.

We can encode neatly the information of the gauge invariance, gauge-fixing and in general the whole structure of the "Hilbert" space by means of the BRST construction. However, even if this may be done at this stage, it is convenient to introduce k Lagrange multipliers Ω_v, which are also treated as new dynamical variables. The new requirement that their associate momenta P_v vanish increases the number of constraints to $2k$. It turns out that having constraints in pairs simplifies the actual treatment in perturbative calculations. What is more important, the additional coordinates associated with Lagrange multipliers can be endowed with an indefinite metric (as in electromagnetism) since the choices in the spurious subspace cannot affect the results in the physical subspace. Thanks to the fact that the constraints come in pairs and to the indefinite metric one can construct a formalism which does not bring in regularisation problems. This is a desirable feature specially in finitely many-body systems where the quantum theory is known and has no such problems.

To implement the BRST "quantization" formalism we proceed by further enlarging the Hilbert space through the introduction of of $2k$ ghost variables η_v, $\bar{\pi}_v$ and their conjugate operators π_v, $\bar{\eta}_v$. The fundamental BRST invariance appears in such an enlarged space. There is a subspace of "physical" states, i.e., of those states which are annihilated by the BRST charge. Moreover, for each physical state of the original problem in the "laboratory" frame there is a state in the subspace of physical states that has non-zero norm which is unique, up to zero-norm states. Moreover, one can prove that the overlap operators in the original Hilbert space are the same as those of "physical" operators in the enlarged space (i.e., those operators mapping physical states into physical states).

The new invariances introduce a large freedom in defining equivalent formulations of the same physical problem. For instance, the enlarged Hilbert space of the original problem in the laboratory frame factorizes into a real subspace $\Psi(q_i')$ and a spurious subspace $\Psi(\phi_v, \Omega_v, \eta_v, \bar{\pi}_v)$. In the description from the transformed frame of reference we treat explicitely the collective coordinates as real degrees of freedom. Thus some combination of the original degrees of freedom q_i become spurious. This trade-off is accomplished by the BRST-gauge fixing mechanism.

Although the formalism has a wider applicability, in these notes we have only explicitely considered cases such that the original hamiltonian has a definite symmetry (which is unrelated to the gauge- or BRST-invariance) and such that the ground state solution does not carry that symmetry. We have shown that the formalism presents an alternative to the usual collective coordinate methods used in such cases. The relation with projection methods extensively used in many-body problems becomes more clear.

In the treatment of approximately broken symmetries the procedure generally follows the same three steps, namely: i) The (classical) determination of the deformed solution of the hamiltonian H. This minimization may include a term with a constant (average) value of the Lagrange multiplier. Typical examples are the BCS approximation for superfluid systems and the cranked-Hartree approximation for deformed nuclei. ii) The construction of the elementary modes of excitation. In the quadratic approximation, the original hamiltonian $\hat{H}$ yields finite-frequency modes plus a zero-energy mode for each direction v along which there is a breakdown of symmetry. In the case of $\hat{H}_{BRST}$, each zero-frequency mode combines with other quadratic terms to yield sistematically the same quadratic (spurious) hamiltonian involving two phonons and two ghosts (cf. the quartet mechanism of Kugo-Ojima [25]). The same (arbitrary) frequency ω_v is associated with the four degrees of freedom, reflecting the supersymmetry of the solution. An indefinite metric is associated with one of the phonons, and all excitations in the spurious sector have positive energy. The vacuum is normalizable, non-degenerate and it is annihilated by the BRST charge in the quadratic approximation. The frequency ω_v cancels (as it should) in the ground state expectation value. iii) Perturbative calculations. To the extent that we remain satisfied with the mean-field approximation and with only the finite frequency modes, the introduction of the BRST formalism has conceptual but not practical consequences. The advantage of $\hat{H}_{BRST}$ over $\hat{H}$ is that the former has not zero-frequency modes (i.e., it does not commute with the generators of the original symmetry group) and thus perturbation theory becomes feasible. In particular, we may treat correctly the effects produced by the fluctuations of the collective coordinates. The price that we must pay is relatively small, since the number of spurious elementary excitations (four per each direction of broken symmetry) is in general much smaller than the number of the original finite-frequency modes.

The requirement of the BRST invariance has become a fundamental tool in the formulation of gauge theories. In these notes we show that it may also be important in the description of mechanical systems from moving frames of reference. In addition, we hope to have conveyed to many-body physicists at least part of the essential physics, as well as some of the aesthetic pleasure, which are present in such formulations.

References

[1] J.L.Borges, *A Personal Anthology*. Grove Press, Inc., New York (1967).

[2] A.Bohr and B.R.Mottelson, *Nuclear Structure, Vol.II.* W.A.Benjamin, Inc., Reading, Massachusetts (1975).

[3] K.Huang, *Quark, Leptons and Gauge Fields*. World Scientific, Singapore (1981).

[4] C.Itsykson and J.B.Zuber, *Quantum Field Theory*. McGraw-Hill, New York (1980).

[5] L.D.Faddeev and V.N.Popov, Phys.Lett., 25B (1967) 29.

[6] A. Hosoya and K. Kikkawa, Nucl. Phys., B101 (1975) 271; I.L.Gervais, A.Jevicki and B.Sakita, Phys. Rep.,23C (1976) 281.

[7] V.Alessandrini, D.R.Bes and B.Machet, Nucl. Phys., B142 (1978) 489.

[8] V.Alessandrini, D.R.Bes and B.Machet, Phys. Lett., 80B (1978) 9; D.R.Bes, G.G.Dussel and R.P.A.Perazzo, Nucl. Phys., A340 (1980) 157; D.R.Bes, O.Civitarese and H.M.Sofia, Nucl. Phys., A370 (1981) 99.

[9] V.Alessandrini, D.R. Bes, O.Civitarese and M.T.Mehr, Phys.Lett., 148B (1984) 395; D.R.Bes, J.Kurchan, M.T.Mehr and G.R.Zemba, Nucl. Phys., A471 (1987) 565.

[10] C.Becchi, A.Rouet and S.Stora, Phys. Lett., 52B (1974) 344; Ann. of Phys., 98 (1976) 278; I.V.Tyutin, Lebedev preprint, FIAN 39 (1979), unpublished; B.L.Voronov and I.V.Tyutin, Theor.Math.Phys. U.S.S.R. 50(1982)218.

[11] E.S.Fradkin and G.A.Vilkovisky, Phys.Lett., 55B (1975) 224; CERN report TH-2332 (1977); I.A.Batalin and G.A.Vilkovisky, Phys.Lett., 69B (1977) 309.

[12] R.Marnelius, Acta Phys. Pol. 1313 (1982) 669.

[13] M.Henneaux, Phys. Rep., 126C (1985) 1.

[14] J.Kurchan, D.R.Bes and S.Cruz Barrios, Phys. Rev., D38 (1988) 3309.

[15] D.R.Bes, S.Cruz Barrios and J.Kurchan, Ann. of Phys., 194 (1989) 227.

[16] J.Kurchan, D.R.Bes and S.Cruz Barrios, Nucl. Phys., A509 (1990) 306.

[17] P.A.M.Dirac, *Lectures in Quantum Mechanics*. Yeshiva University, New York (1964).

[18] E.C.G. Sudarshan and N. Mukunda, *Classical Dynamics. A Modern Perspective.* Wiley, New York (1974).

[19] K. Fujikawa and H. Ui, Prog. Theor. Phys., 75 (1986) 997.

[20] A.R.Edmonds, *Angular Momentum in Quantum Mechanics.* Princeton University Press (1957).

[21] H.Goldstein, *Classical Mechanics.* Addison-Wesley, New York (1959).

[22] K.Bleuler, Helv. Phys. Acta, 23 (1950) 567; S.N.Gupta, Proc. Phys. Soc., A63 (1950) 681.

[23] V.Alessandrini, in *Frontiers in Non-Perturbative Field Theory.* Z. Horwath, L. Palla and A. Patkos, Eds., World Scientific Publ. Co., Singapore (1989); preprint LPTHE 89/08 (1989).

[24] M. Henneaux, in *Quantum Mechanics of Fundamental Systems 1.* C. Teitelboim, Ed., Plenum Press, New York (1988); Ann. of Phys.194 (1989) 281.

[25] T.Kugo and I.Ojima, Phys. Lett. 73B (1978) 459; Suppl. Prog. Theor. Phys. 66 (1979) 1.

[26] F.A.Beresin, *The Method of Second Quantization.* Academic Press, New York (1966).

[27] J.Alfaro and P.H.Damgaard, Ann.of Phys., to be published.

[28] D.Mac Mullan and J. Patterson, University of Glasgow preprints GUTPA/88/2-1, GUTPA/88/3-1.

[29] S.Hwang and R.Marnelius, Nucl. Phys. B315 (1989) 638.

[30] A. Messiah, *Quantum Mechanics.* North-Holland, Amsterdam (1961).

[31] J.P.Blaizot and G.Ripka, *Quantum Theory of Finite Systems.* The MIT Press, Cambridge, Massachusetts (1986).

[32] S.T.Belyaev and V.G.Zelevinski, Nucl. Phys. 39 (1962) 583; E.R.Marshalek, Nucl. Phys. A224 (1974) 221, 245.

[33] D.R.Bes, G.G.Dussel, R.A.Broglia, B.R.Mottelson and R.Liotta, Phys. Lett. 52B (1974) 253; D.R.Bes, Prog. Theor. Phys. Suppl 74 (1983) 1.

[34] D.R.Bes and R.A.Broglia, in *Elementary Modes of Excitation in Nuclei.* A.Bohr and R.A.Broglia, Eds., Academic Press, New York (1977).

[35] E.R.Marshalek and J.Weneser, Phys. Rev., C2 (1970) 1682.

[36] D.J.Thouless and J.G.Valatin, Nucl. Phys., 31 (1962) 211.

[37] D.R.Inglis, Phys. Rev., 96 (1954) 1059; 97 (1955) 701.

[38] M.Harvey, Advan. Nucl. Phys., 1 (1968) 67.

[39] D.R.Bes, R. De Luca and J.Kurchan, to be published.

[40] E.R.Marshalek, Nucl. Phys., A275 (1977) 416; Nucl. Phys. A381 (1982) 240.

[41] D.Janssen and I.N.Mikhailov, Nucl. Phys.,A318 (1979) 390; V.G.Zelevinski, Nucl. Phys., A344 (1980) 109; J.L.Egido, H.J.Mang and P.Ring, Nucl. Phys. A341 (1980) 229.